Photographic Atlas
of
Anatomical Models

WILLIAM BROTHERS
SAN DIEGO MESA COLLEGE
SAN DIEGO, CALIFORNIA
1/2015

The *Photographic Atlas of Anatomical Models* is designed to allow students to focus on the most frequently learned structures in an introductory human anatomy course. This atlas is to supplement the resources normally used in human anatomy and to allow students to have a convenient book to review. The models chosen for the atlas are used in the human anatomy and human anatomy and physiology classes taught at San Diego Mesa College. The atlas is intended for Mesa's students, but can be utilized by many courses, programs and universities. The photographs are labeled with letters and or numbers with an identification key listed nearby. The intent of this format is for observing the structures but also self-testing by conveniently covering the identification key. Bones are not included in the atlas since most textbooks have excellent drawings and photographs of the skeletal system.

The atlas is organized by body systems. The first page or pages of each chapter has thumbnail photos of the models used for each system. The manufacturer of the models is included to aid the user of the atlas when studying the models and for schools who want to purchase them from a variety of distributors.

The manufacturers of the models shown in this atlas are: 3B Scientific, Denoyer-Geppert International, GPI Anatomicals, Laerdal Medical, and Somso Modelle. These names and models are Trademarked and Copyrighted.

The terminology of the anatomical structures utilized in the atlas is those endorsed by the International Federation of Associations of Anatomists as published in the *Terminologies Anatomica*. Listed below are some structures and their commonly used synonyms.

Anterior interventricular branch	Left anterior descending artery
Auditory tube	Eustachian tube
Auricle	Pinna
Brachiocephalic trunk	Innominate artery
Bulbourethral glands	Cowper's glands
Celiac trunk	Celiac artery
Ductus deferens	Vas deferens
Fibular (Fibularis)	Peroneal (Peroneus)
Glomerular capsule	Bowman's capsule
Median cubital vein	Median basilic vein
Patellar ligament	Patellar tendon
Posterior interventricular branch	Posterior descending artery
Pulmonary trunk	Pulmonary artery
Subendocardial branches	Purkinje Fibers
Uterine tube	Fallopian tube

Sincerely, William Brothers
Please send any comments or corrections to bbrother@sdccd.edu

Table of Contents

Integumentary System Models 5
Skin list 6
Skin model 7

Skeletal System Models 8
Bone tissue 9
Beauchene skull 10
Shoulder joint 12
Elbow joint 13
Hip joint 14
Knee joint 15

Muscular System Models 16
Male muscular figure 18
Muscle man 25
Torso (3B) 31
Torso (Somso) 36
Torso (Denoyer-Geppert) 39
Head and neck 42
Shoulder muscles 43
Hip muscles 44
Upper extremity 45
Lower extremity 50
Shoulder section 54
Elbow section 55
Hip section 56
Muscle fiber 57

Nervous System Models 58
Brain 60
Brain sagittal section 64
Head and neck 66
Brain axial sections 67
Ventricles 74
Neuron 75
Spinal cord in fifth vertebra 76
Spinal cord in spinal canal 78
Nervous system (flat models) 80
Eye 82
Ear 84
Inner ear labyrinth 85
Cochlea 86

Cardiovascular System Models 87
Male muscular figure 89

Torso (3B) 90
Torso (Somso) 92
Circulatory system (flat) 94
Circulatory system (wire) 98
Blood vessels (3B) 103
Heart (Denoyer-Geppert) 104
Heart (Laerdal Medical) 107
Heart (Somso) 110

Respiratory System Models 112
Male muscular figure 114
Torso (Somso) 115
Nose 116
Head and neck (Denoyer-Geppert) 117
Head and neck (3B) 119
Larynx 120
Lungs 121
Larynx and trachea 122
Alveoli 123

Digestive System Models 124
Male muscular figure 125
Torso (3B) 126
Torso (Somso) 129
Stomach 132
Digestive system (flat) 133
Head and neck (Denoyer-Geppert) 136
Head and neck (3B) 137
Teeth 138
Pancreas, gall bladder and duodenum 140
Small intestinal with villi 141

Urinary System Models 142
Male muscular figure 143
Torso (3B) 144
Torso (Somso) 145
Kidney 146
Kidney nephron 147
Renal corpuscle 148

Reproductive System Models 149
Male reproductive system (Somso) 150
Male reproductive system (Carolina Biological) 152
Male reproductive system (Denoyer-Geppert) 153
Male reproductive system Torso (Somso) 154
Female reproductive system Torso (Somso) 155
Female reproductive system (Somso) 156
Female reproductive system frontal view (Somso) 158
Female reproductive system (Bobbitt) 160

Chapter 1
Integumentary System Model

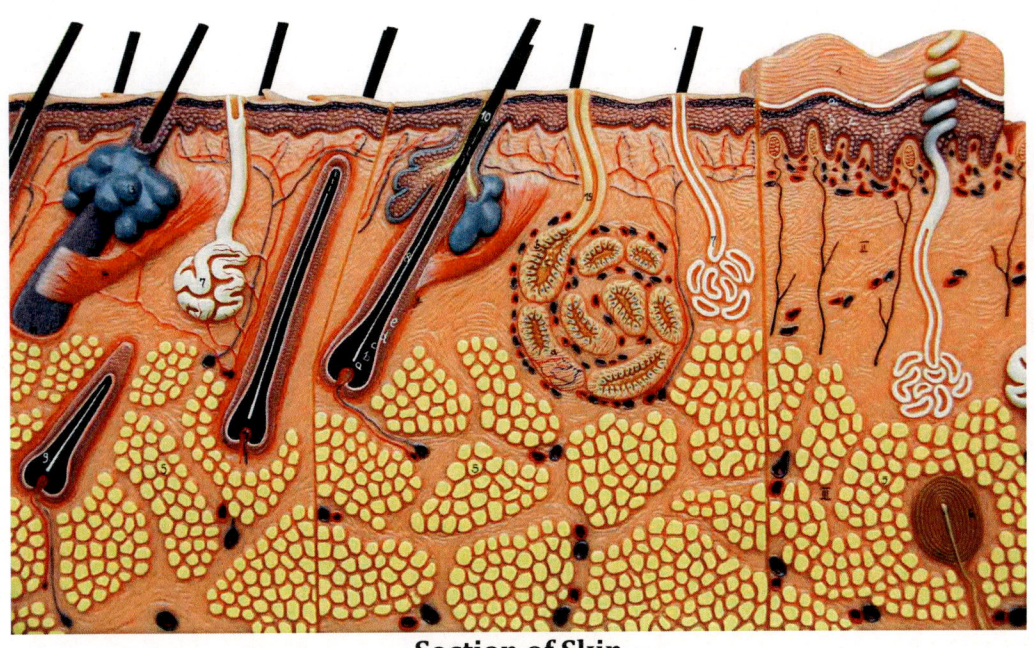

Section of Skin
(Copyright Somso)

Identification Key for Figures 1-1 and 1-2

    A. Skin from scalp
    B. Skin from axilla
    C. Skin from palm or sole
    D. Epidermis
    E. Dermis
    F. Hypodermis

1. Stratum corneum
2. Stratum lucidum
3. Stratum granulosum
4. Stratum spinosum
5. Stratum germinativum (basale)
6. Dermal papilla
7. Tactile (Meissner's) corpuscle
8. Sweat duct
9. Sweat gland
10. Lamellated (Pacinian) corpuscle
11. Sebaceous gland
12. Arrector pili muscle
13. Hair follicle sheath
14. Hair papilla
15. Hair bulb
16. Hair root
17. Hair shaft
18. Apocrine sweat gland

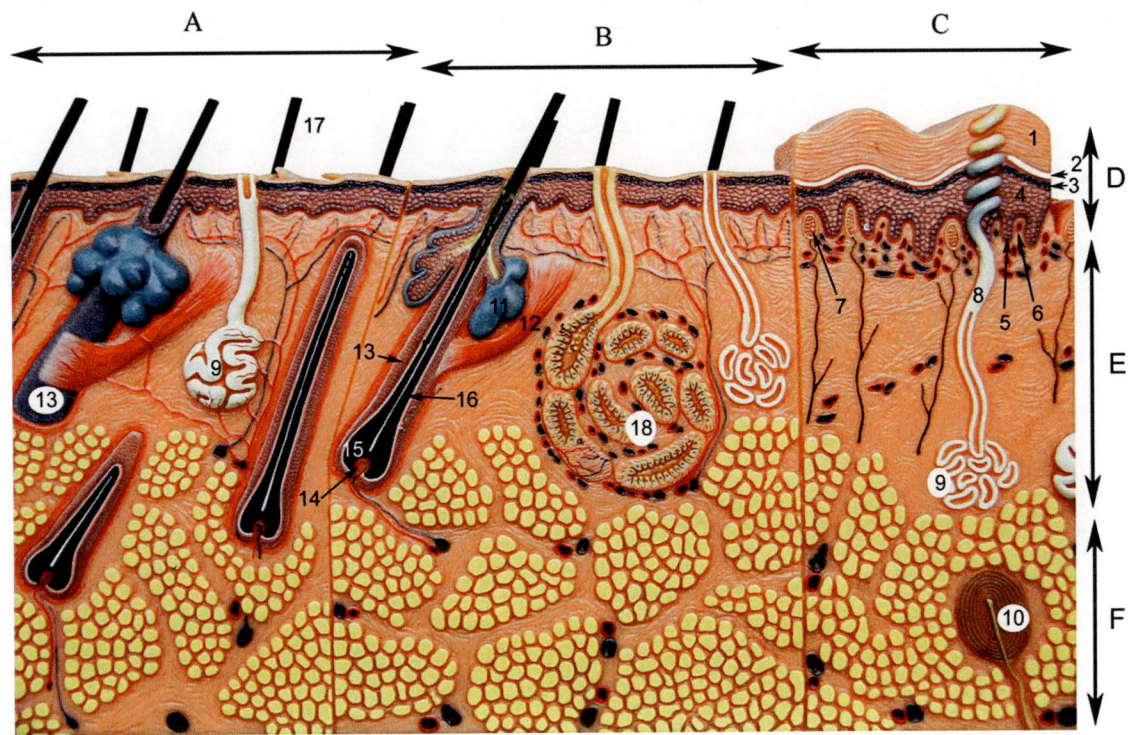

Figure 1-1. Three Sections of Skin
Copyright Somso

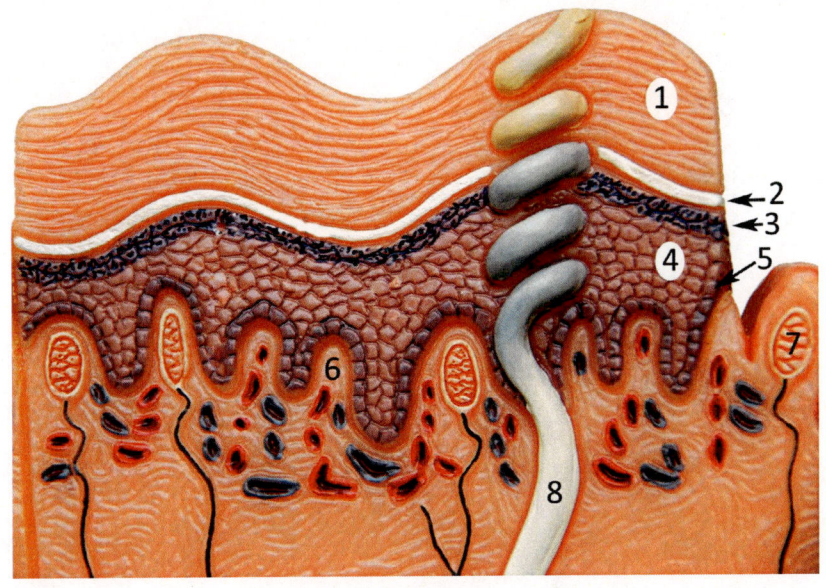

Figure 1-2. Skin – palm
Copyright Somso

Chapter 2
Skeletal System Models

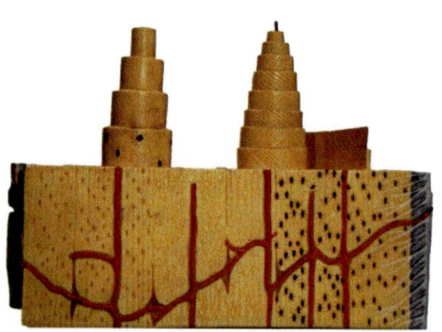

Bone Tissue
(Copyright Somso)

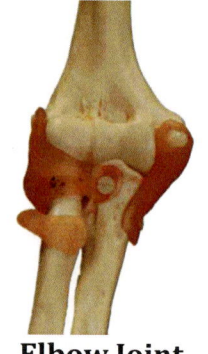

Elbow Joint
(Copyright Somso)

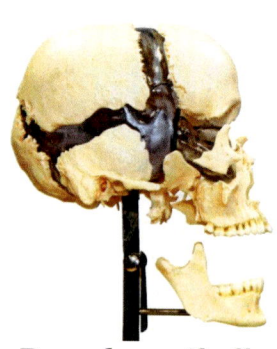

Beauchene Skull
(Copyright Somso)

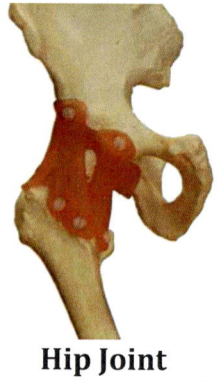

Hip Joint
(Copyright Somso)

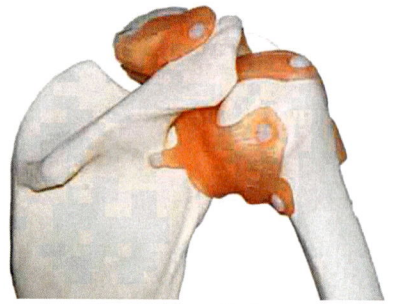

Shoulder Joint
(Copyright 3B Scientific)

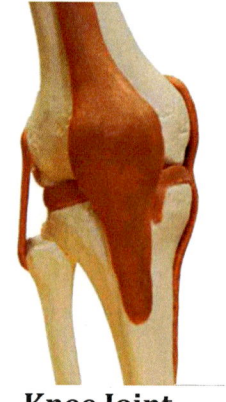

Knee Joint
(Copyright Somso)

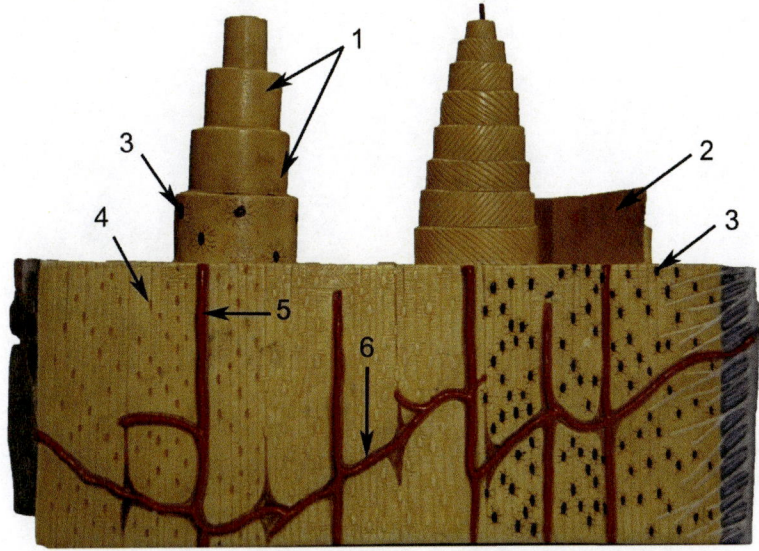

Figure 2-1. Bone Tissue – longitudinal section
Copyright Somso

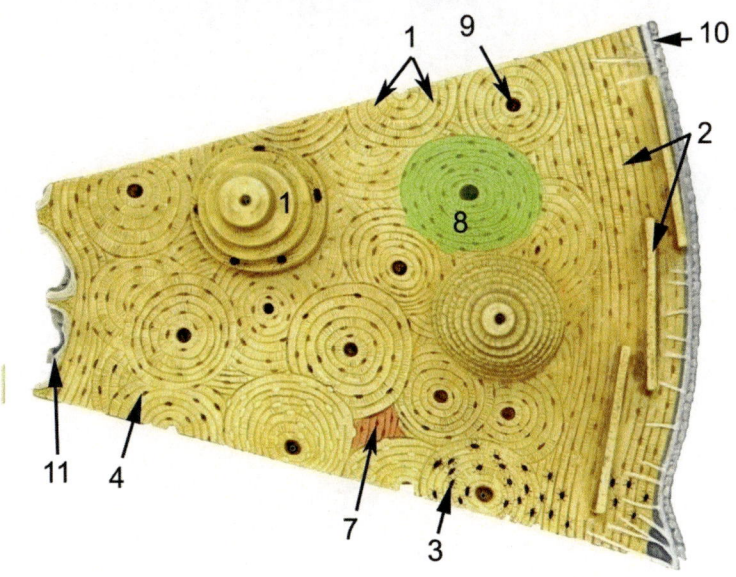

Figure 2-2. Bone Tissue – cross section
Copyright Somso

1. Concentric lamellae
2. Circumferential lamellae
3. Osteocyte
4. Lacuna
5. Blood vessel in central canal
6. Blood vessel in perforating canal
7. Interstitial lamellae
8. Osteon (Haversian system)
9. Central canal (Haversian canal)
10. Periosteum
11. Endosteum

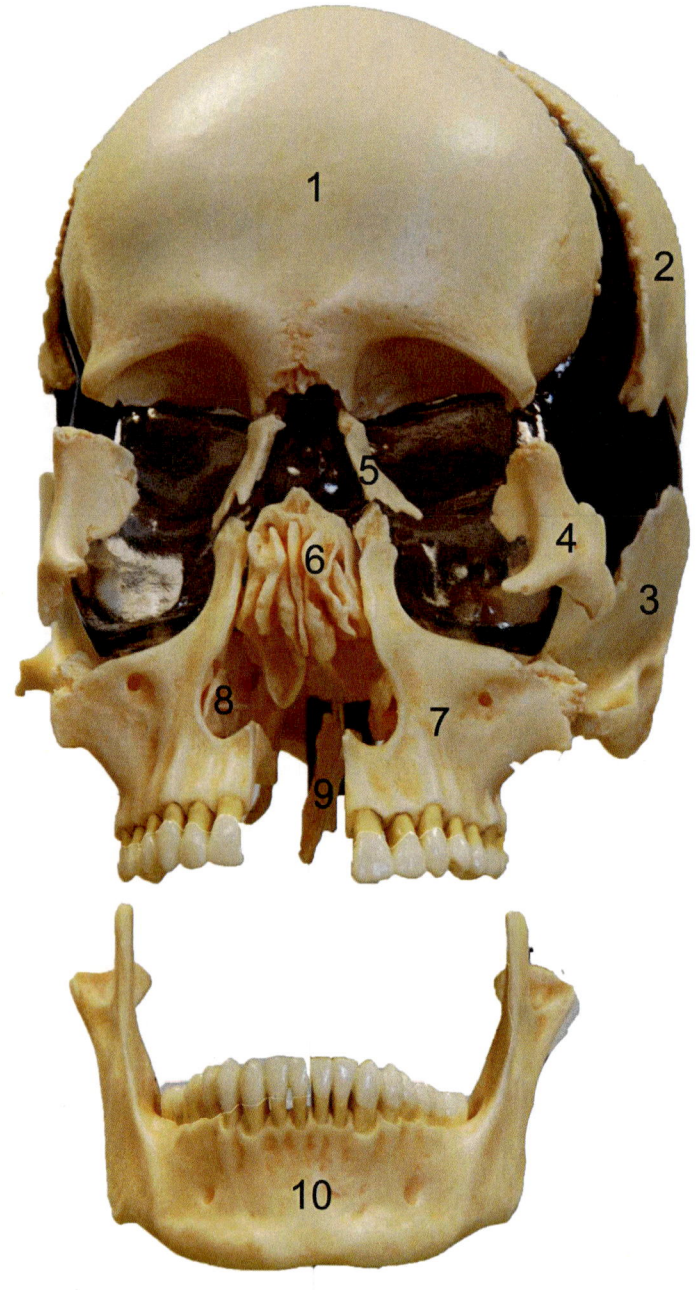

Figure 2-3. Beauchene Skull – anterior
Copyright Somso

1. Frontal
2. Parietal
3. Temporal
4. Zygomatic
5. Nasal
6. Ethmoid
7. Maxilla
8. Inferior nasal concha
9. Vomer
10. Mandible

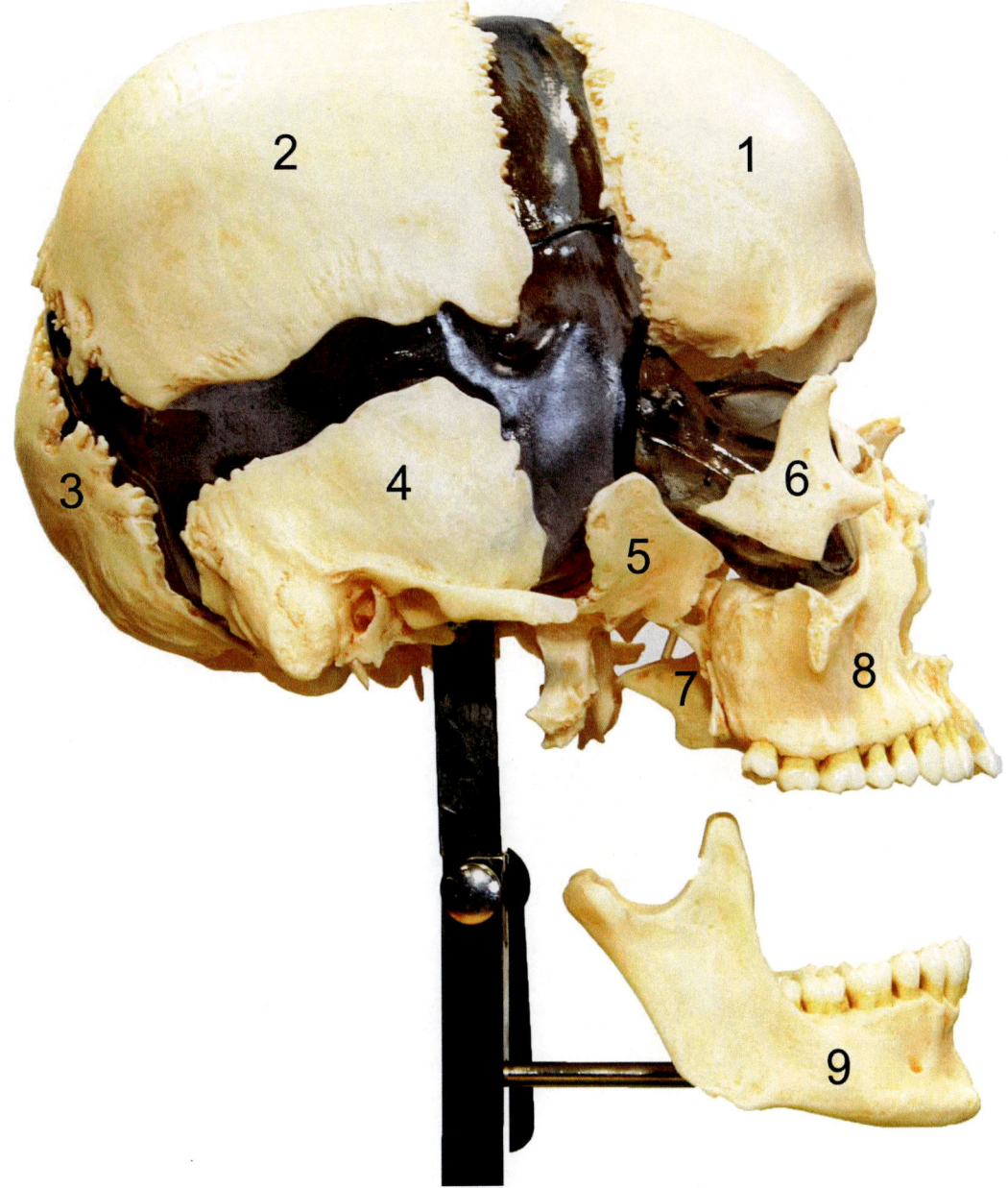

Figure 2-4. Beauchene Skull - lateral
Copyright Somso

1. Frontal
2. Parietal
3. Occipital
4. Temporal
5. Sphenoid
6. Zygomatic
7. Vomer
8. Maxilla
9. Mandible

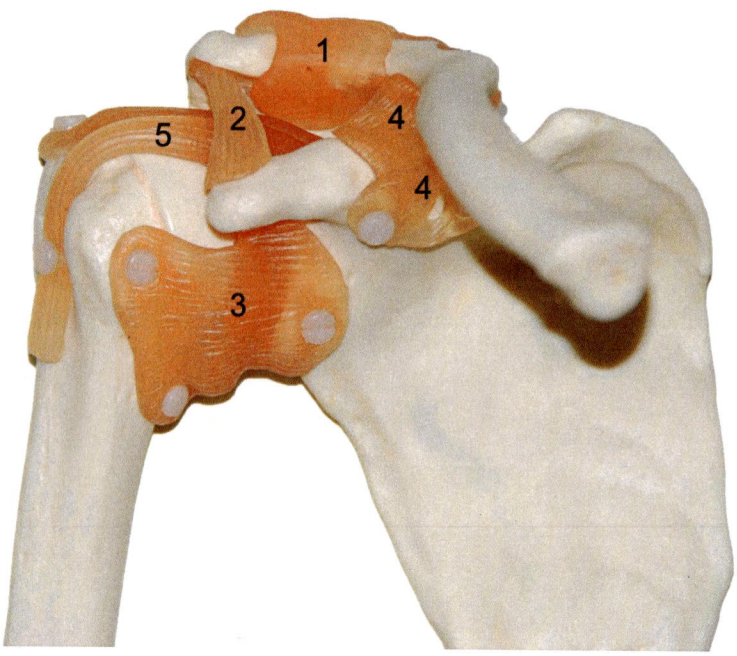

Figure 2-5. Shoulder Joint - anterior
Copyright 3B Scientific

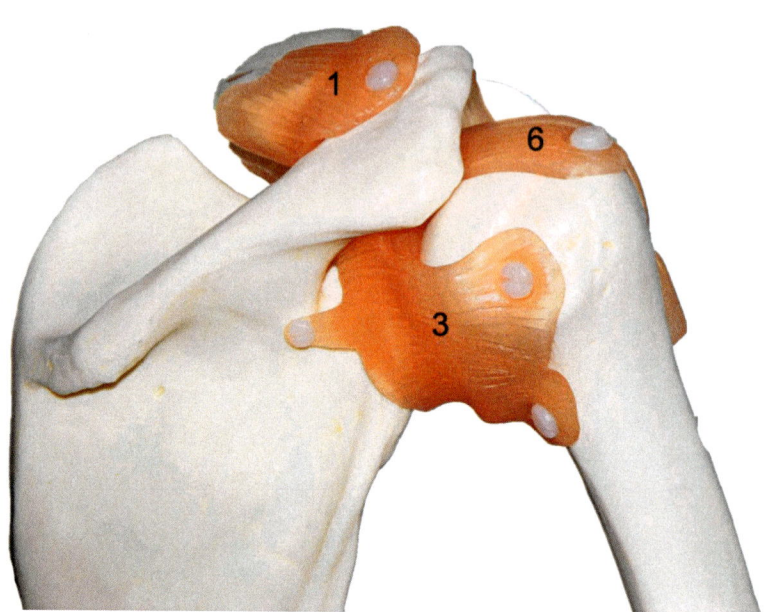

Figure 2-6. Shoulder Joint - posterior
Copyright 3B Scientific

1. Acromioclavicular ligament
2. Coracoacromial ligament
3. Glenohumeral ligament
4. Coracoclavicular ligament
5. Biceps brachii tendon
6. Superior glenohumeral ligament

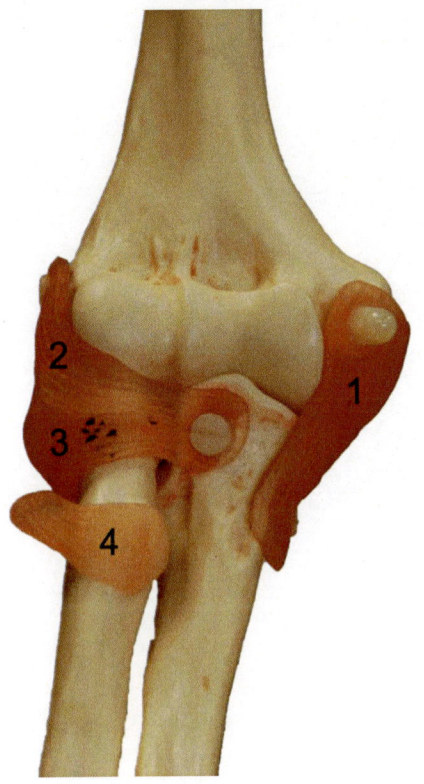

Figure 2-7. Elbow Joint – anterior
Copyright Somso

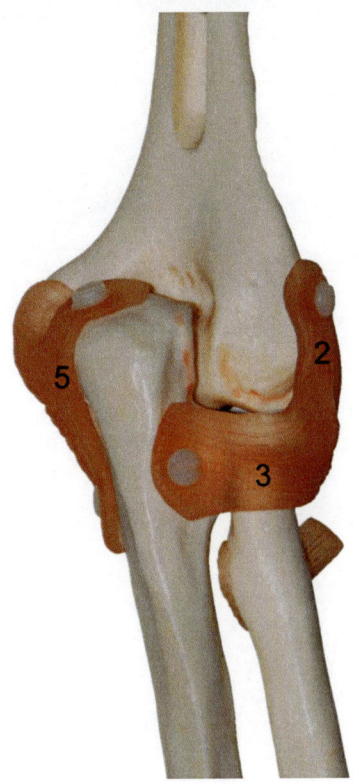

Figure 2-8. Elbow Joint - posterior
Copyright Somso

1. Ulnar collateral ligament (anterior band)
2. Radial collateral ligament
3. Annular ligament of radius
4. Biceps brachii tendon
5. Ulnar collateral ligament (transverse band)

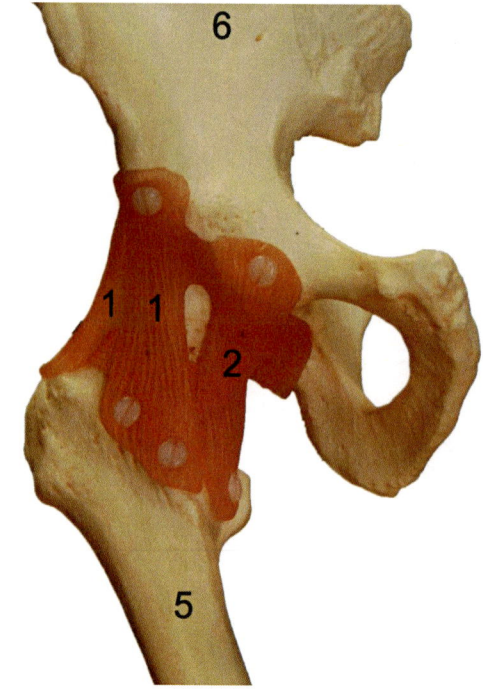

Figure 2-9. Hip Joint - anterior
Copyright Somso

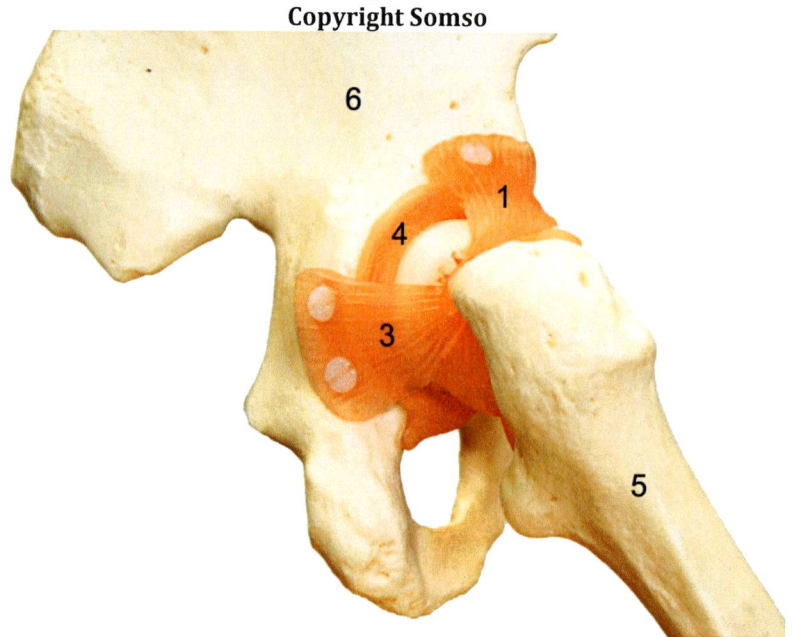

Figure 2-10. Hip Joint - lateral
Copyright Somso

1. Iliofemoral ligament
2. Pubofemoral ligament
3. Ischiofemoral ligament
4. Acetabular labrum
5. Femur
6. Os coxae

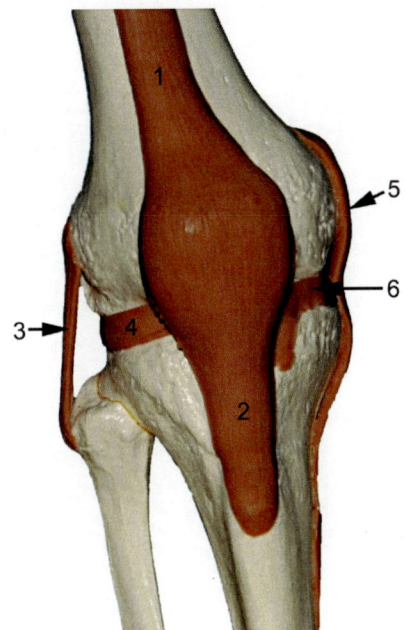

Figure 2.11. Knee Joint - anterior

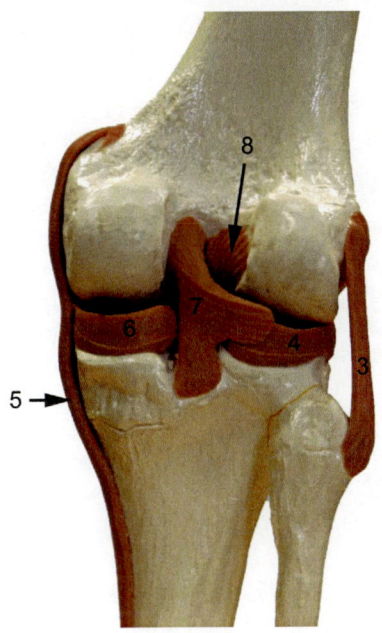

Figure 2.12. Knee Joint - posterior

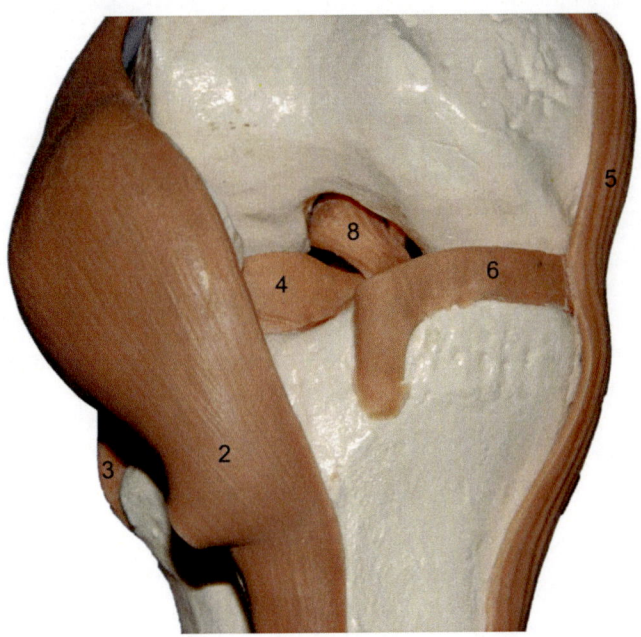

Figure 2-13. Knee - anterior (patella removed)
Copyright Somso

1. Quadriceps tendon
2. Patellar ligament
3. Fibular (lateral) collateral ligament
4. Lateral meniscus
5. Tibial (medial) collateral ligament
6. Medial meniscus
7. Posterior cruciate ligament
8. Anterior cruciate ligament

Chapter 3
Muscular System Models

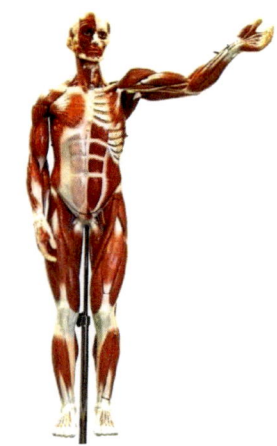

Male Muscle Figure
(Copyright Somso)

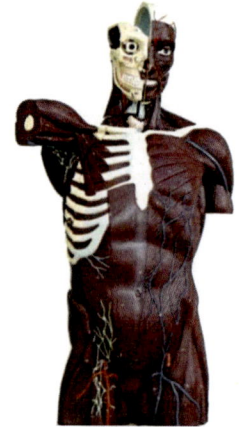

Torso
(Copyright Somso)

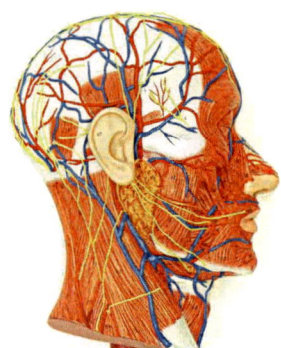

Half Head with Musculature
(Copyright 3B Scientific)

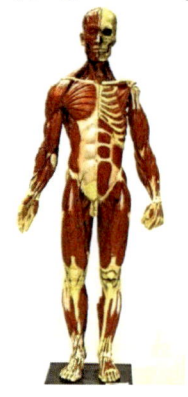

Muscle Man
(Copyright 3B Scientific)

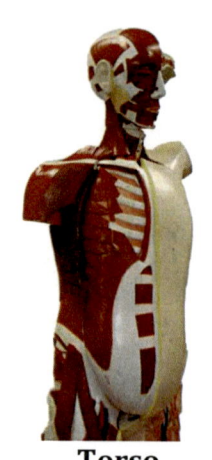

Torso
(Copyright Denoyer-Geppert)

Rotator Cuff
(Copyright GPI Anatomicals)

Torso
(Copyright 3B Scientific)

Hip Muscles
(Copyright GPI Anatomicals)

Upper Extremity
(Copyright Somso)

Lower Extremity
(Copyright Somso)

Muscle Fiber
(Copyright Somso)

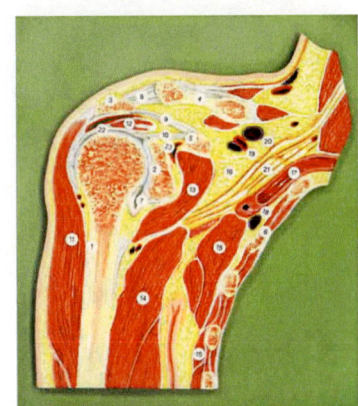

Shoulder Section
(Copyright Somso)

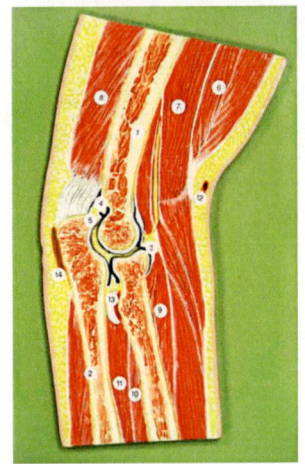

Elbow Section
(Copyright Somso)

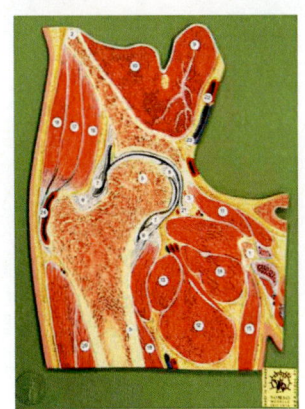

Hip Section
(Copyright Somso)

Male Muscle Figure (Somso)

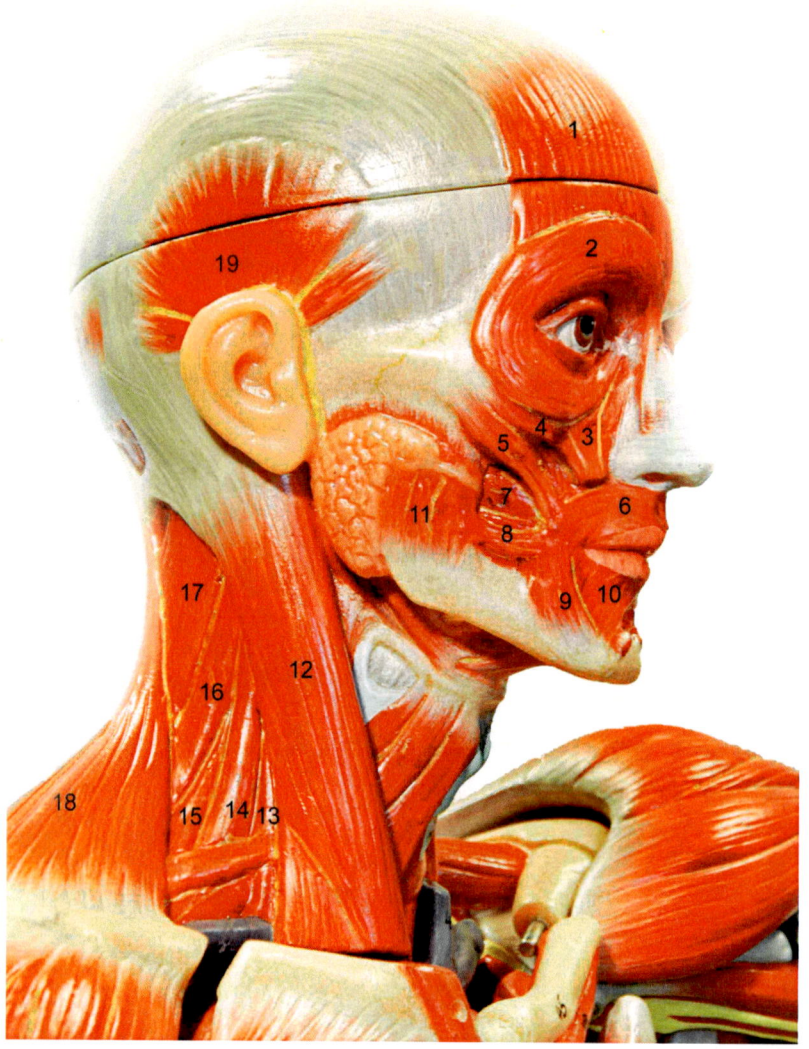

Figure 3-1. Head and Neck - lateral superficial muscles
(Copyright Somso)

1. Frontal belly of occipitofrontalis
2. Orbicularis oculi
3. Levator labii superioris
4. Zygomaticus minor
5. Zygomaticus major
6. Orbicularis oris
7. Buccinator
8. Risorius
9. Depressor anguli oris
10. Depressor labii inferioris
11. Masseter
12. Sternocleidomastoid
13. Anterior scalene
14. Middle scalene
15. Posterior scalene
16. Levator scapulae
17. Splenius capitis
18. Trapezius
19. Auricularis

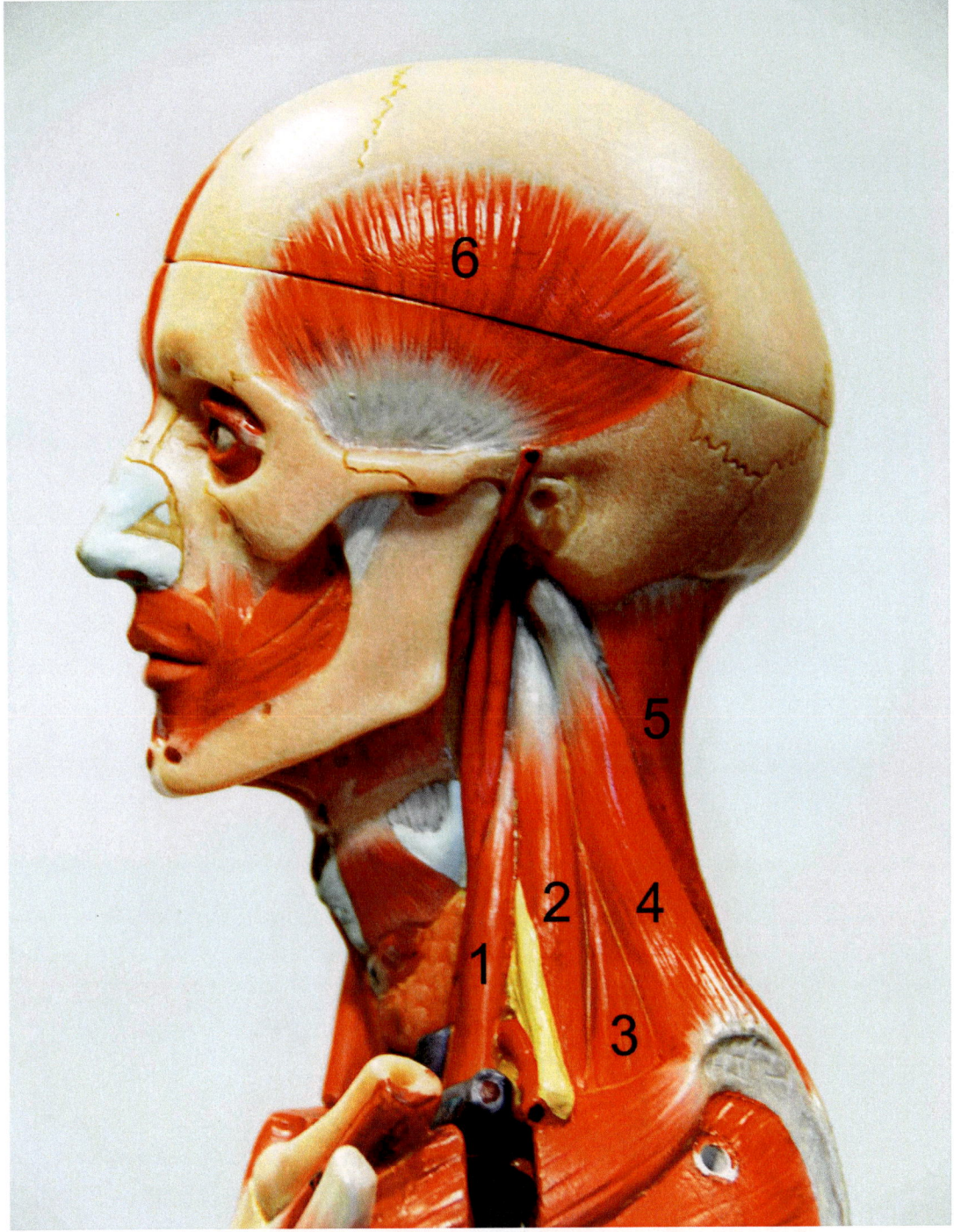

Figure 3-2. Head and Neck - lateral deep muscles
(Copyright Somso)

1. Anterior scalene
2. Middle scalene
3. Posterior scalene
4. Levator scapulae
5. Splenius capitis
6. Temporalis

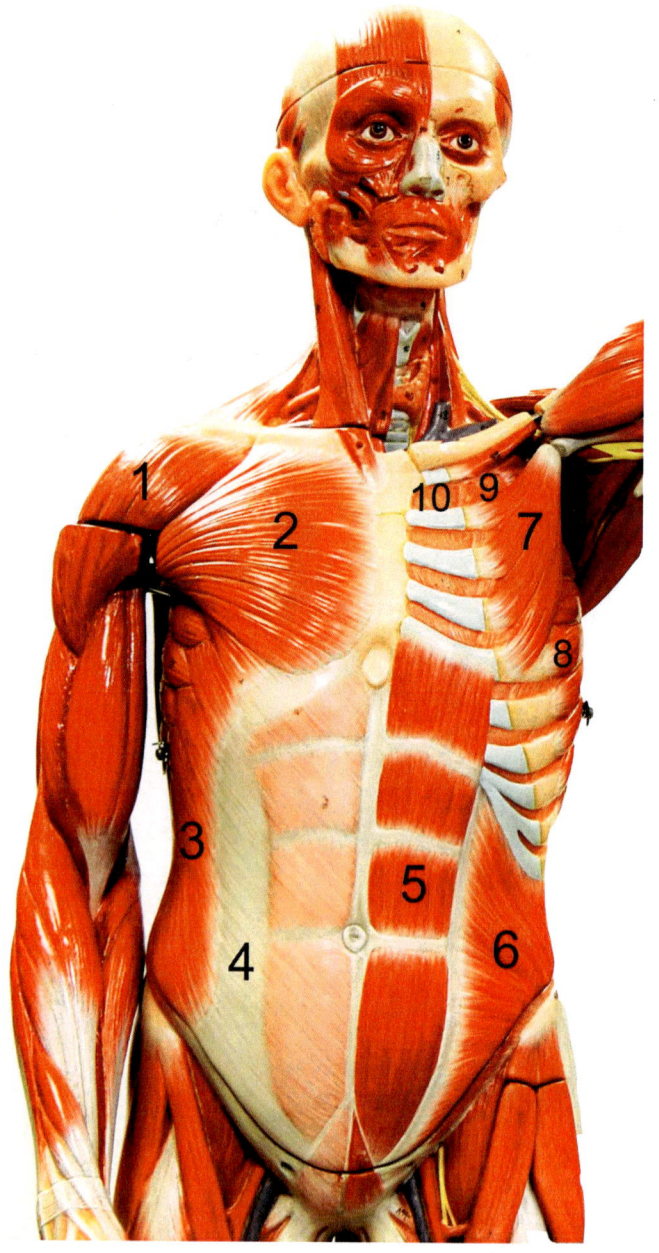

Figure 3-3. Trunk - anterior
(Copyright Somso)

1. Deltoid
2. Pectoralis major
3. External oblique
4. Aponeurosis of external oblique
5. Rectus abdominis
6. Internal oblique
7. Pectoralis minor
8. Serratus anterior
9. External intercostal
10. Internal intercostal

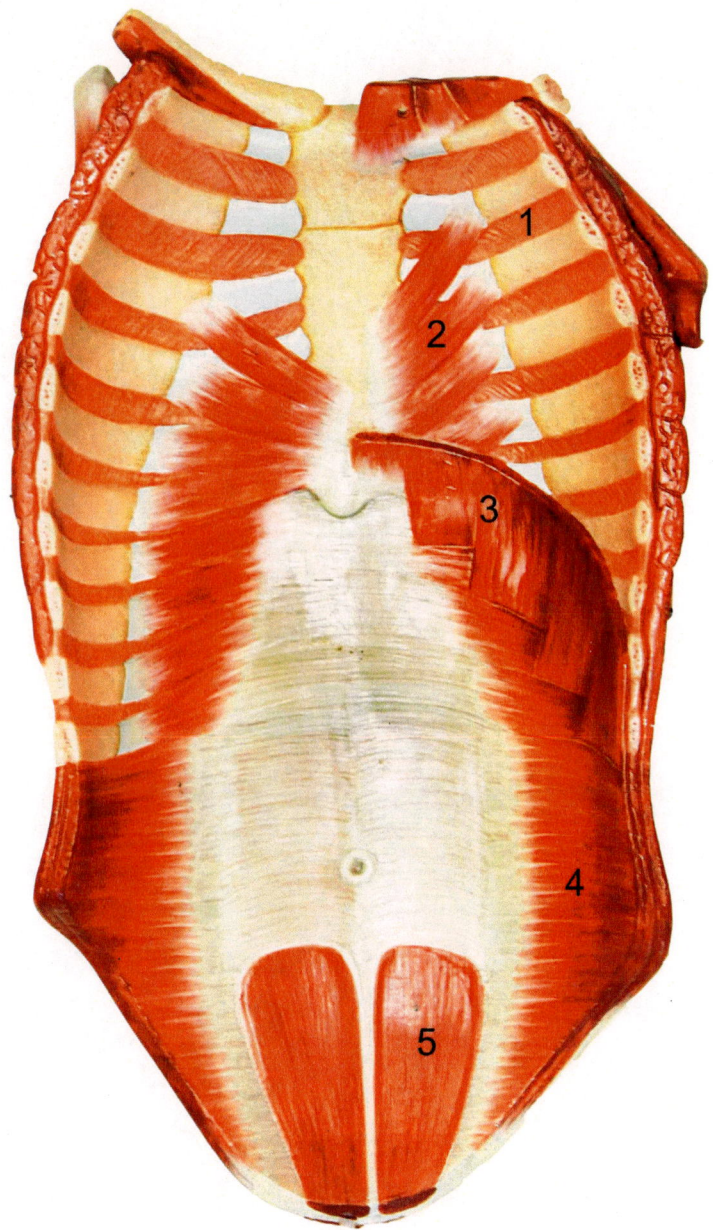

Figure 3-4. Trunk – internal trunk
(Copyright Somso)

1. Internal intercostal
2. Transversus thoracis
3. Diaphragm (cut)
4. Transversus abdominis
5. Rectus abdominis

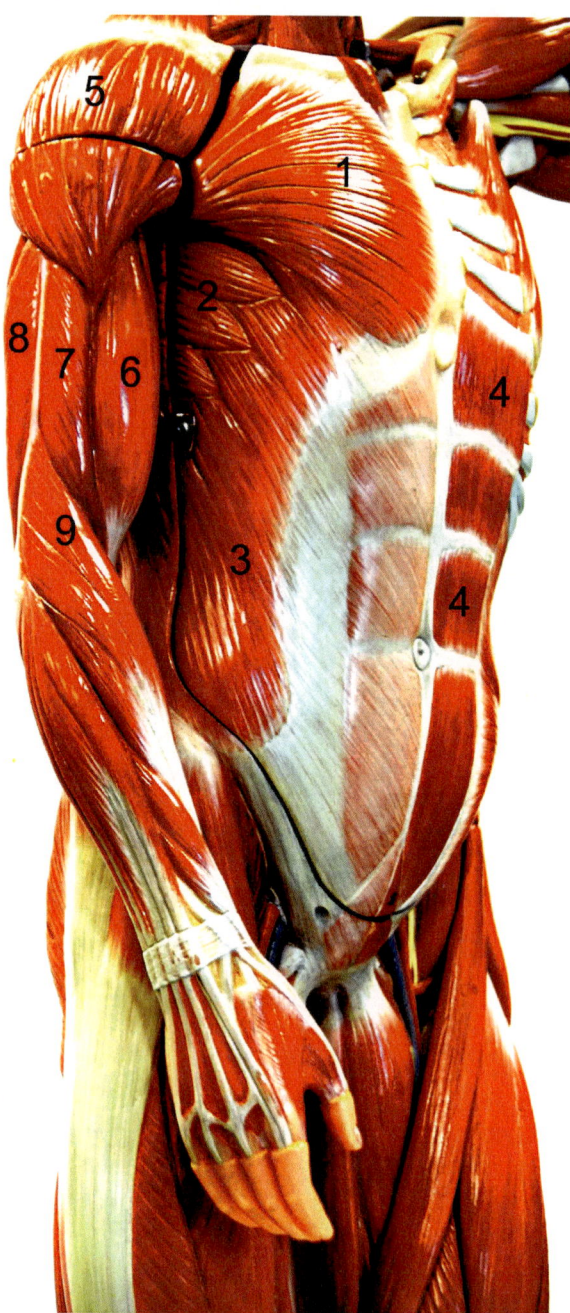

Figure 3-5. Trunk - lateral
(Copyright Somso)

1. Pectoralis major
2. Serratus anterior
3. External oblique
4. Rectus abdominis
5. Deltoid
6. Biceps brachii long head
7. Brachialis
8. Triceps brachii lateral head
9. Brachioradialis

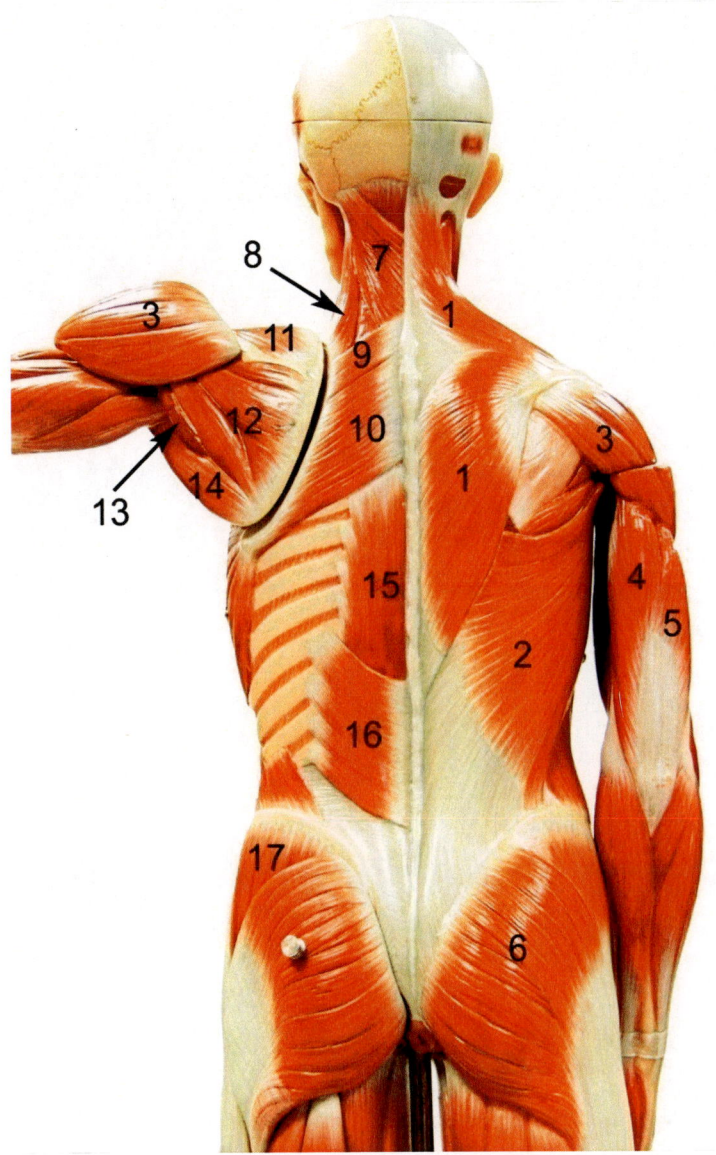

Figure 3-6. Trunk – posterior
(Copyright Somso)

1. Trapezius
2. Latissimus dorsi
3. Deltoid
4. Triceps brachii long head
5. Triceps brachii lateral head
6. Gluteus maximus
7. Splenius capitis
8. Levator scapulae
9. Rhomboid minor
10. Rhomboid major
11. Supraspinatus
12. Infraspinatus
13. Teres minor
14. Teres major
15. Erector spinae
16. Serratus posterior inferior
17. Gluteus medius

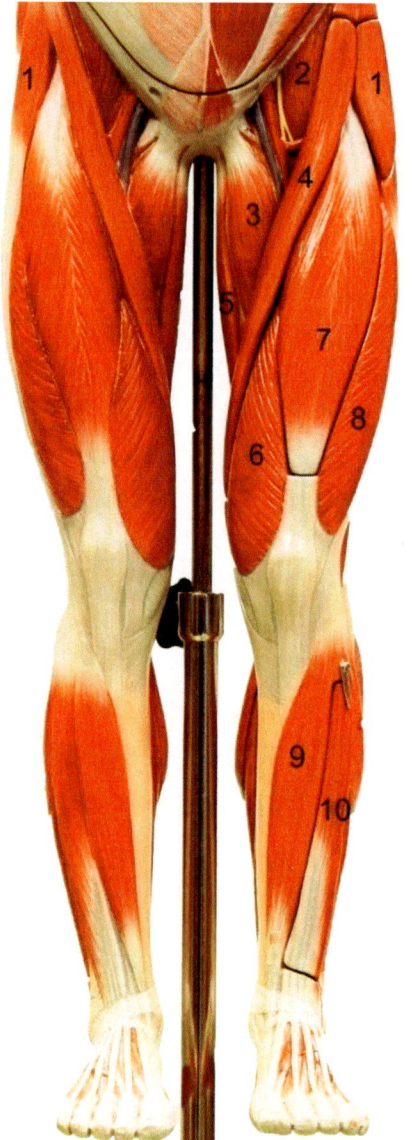
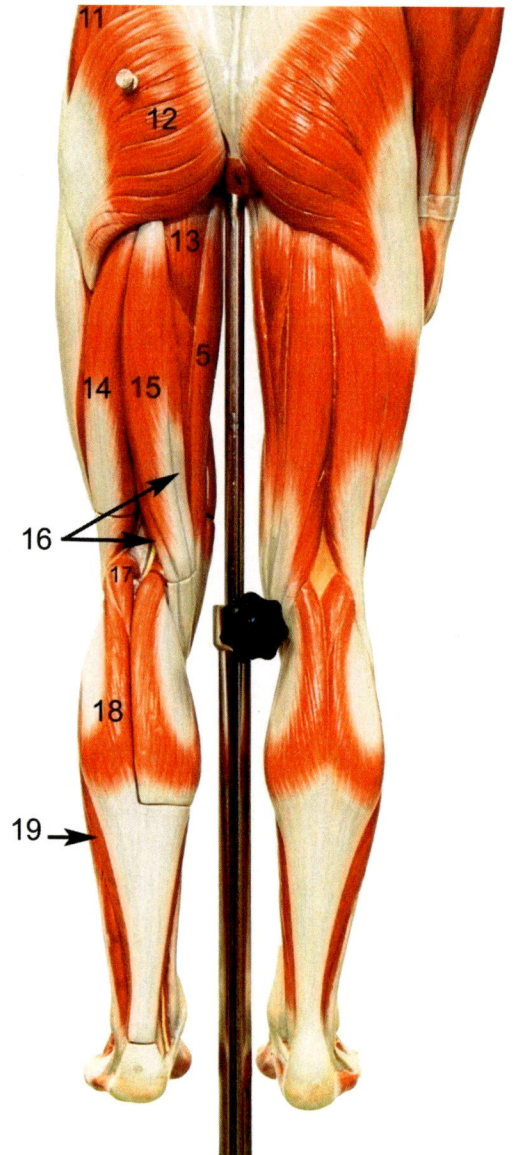

Figure 3-7. Lower Extremity - anterior
(Copyright Somso)
1. Tensor fasciae latae
2. Iliopsoas
3. Adductor longus
4. Sartorius
5. Gracilis
6. Vastus medialis
7. Rectus femoris
8. Vastus lateralis
9. Tibialis anterior
10. Extensor digitorum longus

Figure 3-8. Lower Extremity - posterior
(Copyright Somso)
11. Gluteus medius
12. Gluteus maximus
13. Adductor magnus
14. Biceps femoris long head
15. Semitendinosus
16. Semimembranous
17. Plantaris
18. Gastrocnemius
19. Soleus

Muscle Man

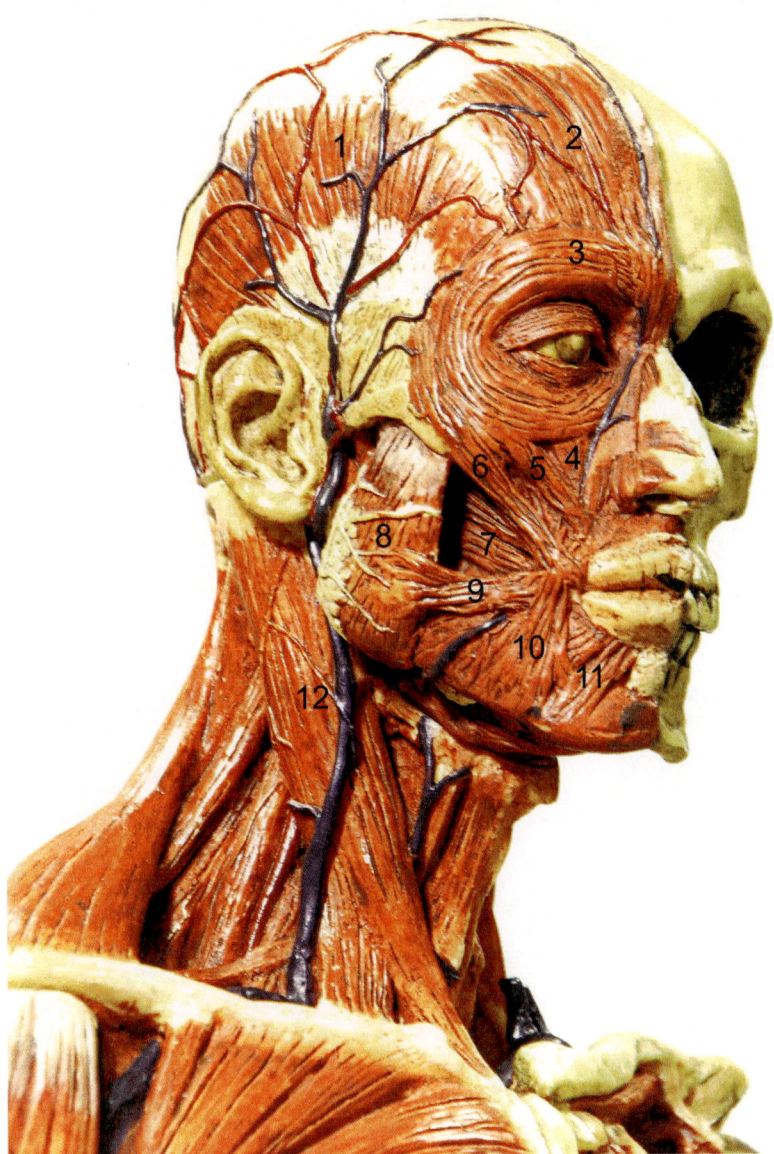

Figure 3-9. Head and Neck - superficial
(Copyright 3B Scientific)

1. Temporalis
2. Frontal belly of occipitofrontalis
3. Orbicularis oculi
4. Levator labii superioris
5. Zygomaticus minor
6. Zygomaticus major
7. Buccinator
8. Masseter
9. Risorius
10. Depressor anguli oris
11. Depressor labii inferioris
12. Sternocleidomastoid

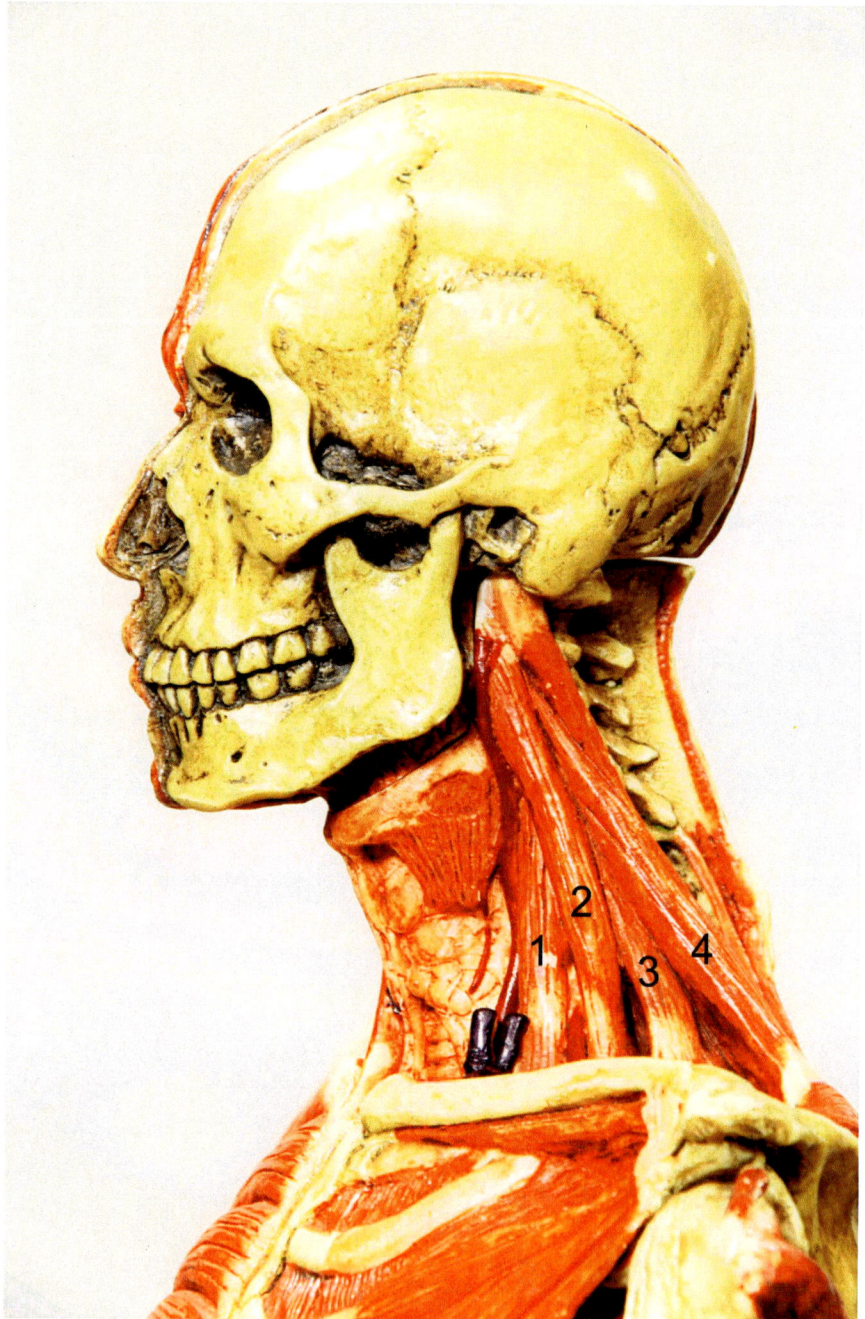

Figure 3-10. Head and Neck - deep
(Copyright 3B Scientific)

1. Anterior scalene
2. Middle scalene
3. Posterior scalene
4. Levator scapulae

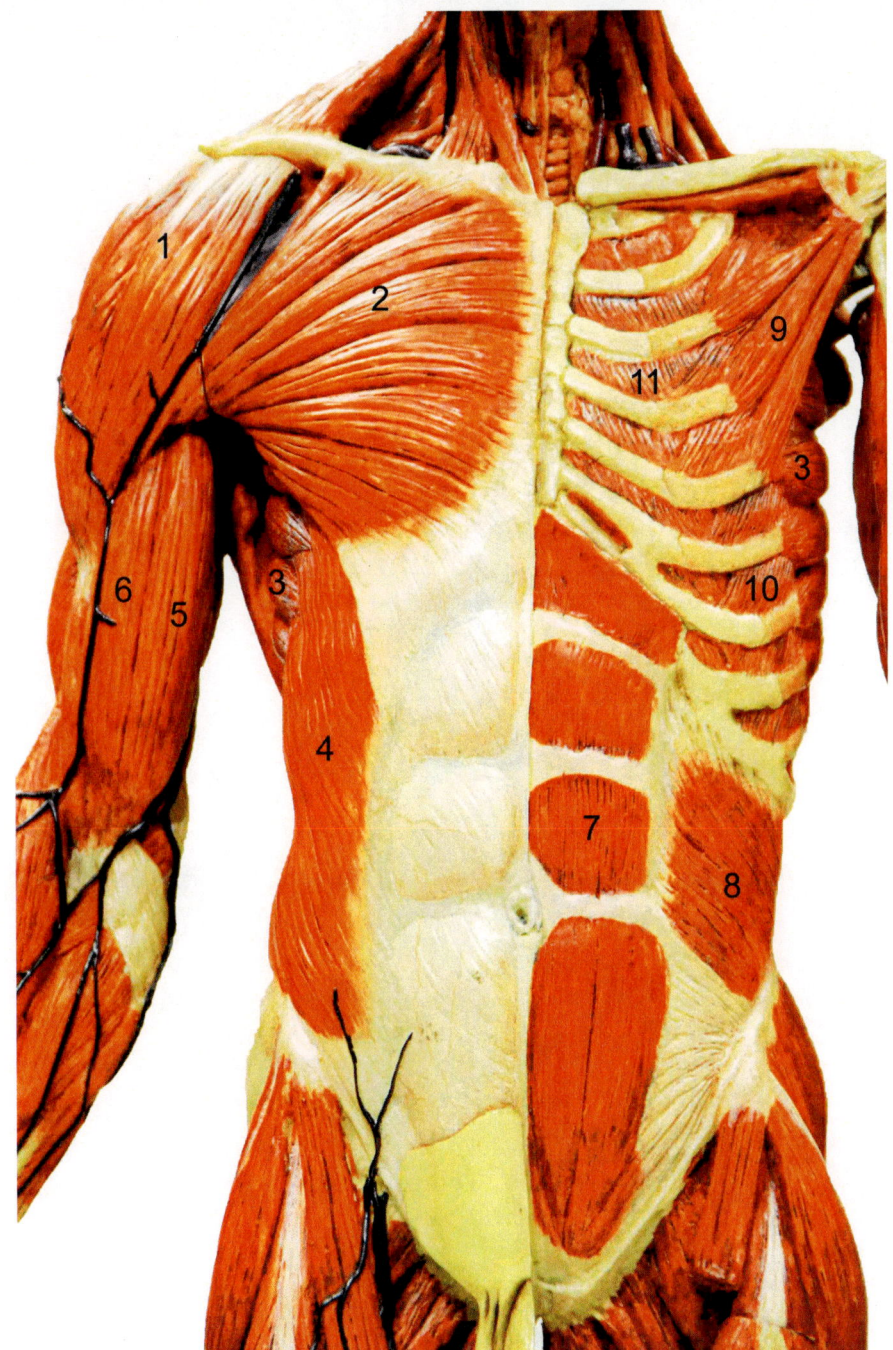

Figure 3-11. Trunk - anterior
(Copyright 3B Scientific)

1. Deltoid
2. Pectoralis major
3. Serratus anterior
4. External oblique
5. Biceps brachii short head
6. Biceps brachii long head
7. Rectus abdominis
8. Internal oblique
9. Pectoralis minor
10. External intercostal
11. Internal intercostal

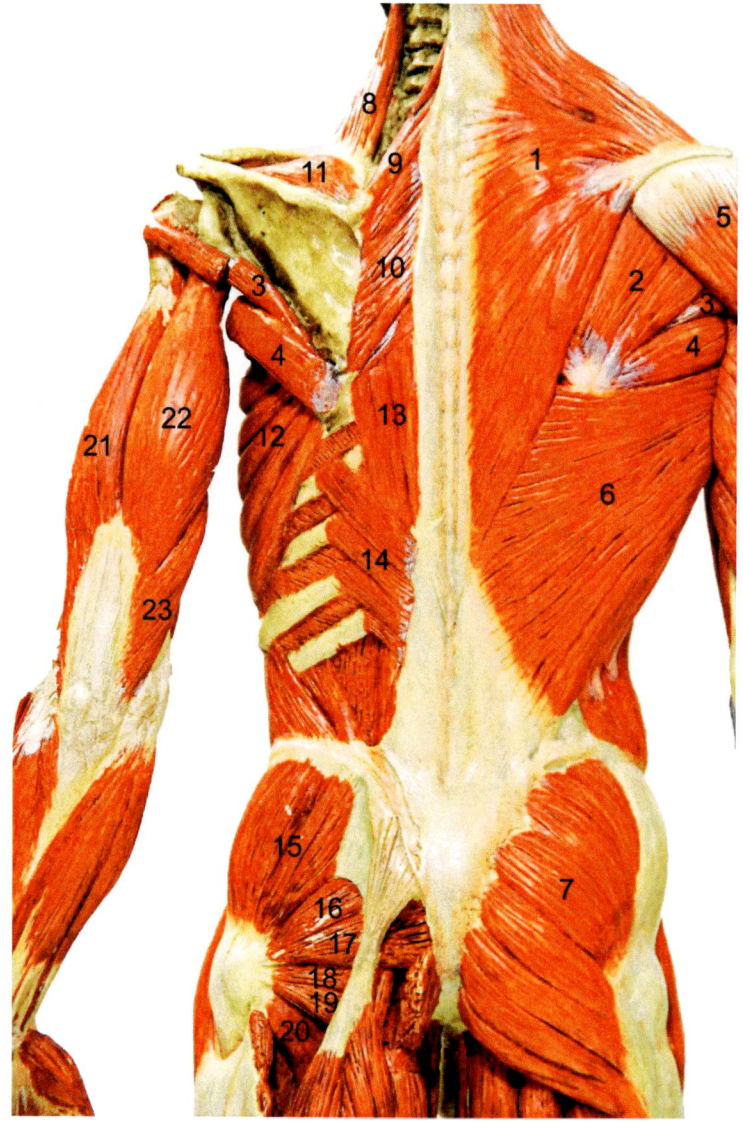

Figure 3-12. Trunk - posterior
(Copyright 3B Scientific)

1. Trapezius
2. Infraspinatus
3. Teres minor
4. Teres major
5. Deltoid
6. Latissimus dorsi
7. Gluteus maximus
8. Levator scapulae
9. Rhomboid minor
10. Rhomboid major
11. Supraspinatus
12. Serratus anterior
13. Erector spinae
14. Serratus posterior inferior
15. Gluteus medius
16. Piriformis
17. Superior gemellus
18. Obturator internus
19. Inferior gemellus
20. Quadratus femoris
21. Triceps brachii lateral head
22. Triceps brachii long head
23. Triceps brachii medial head

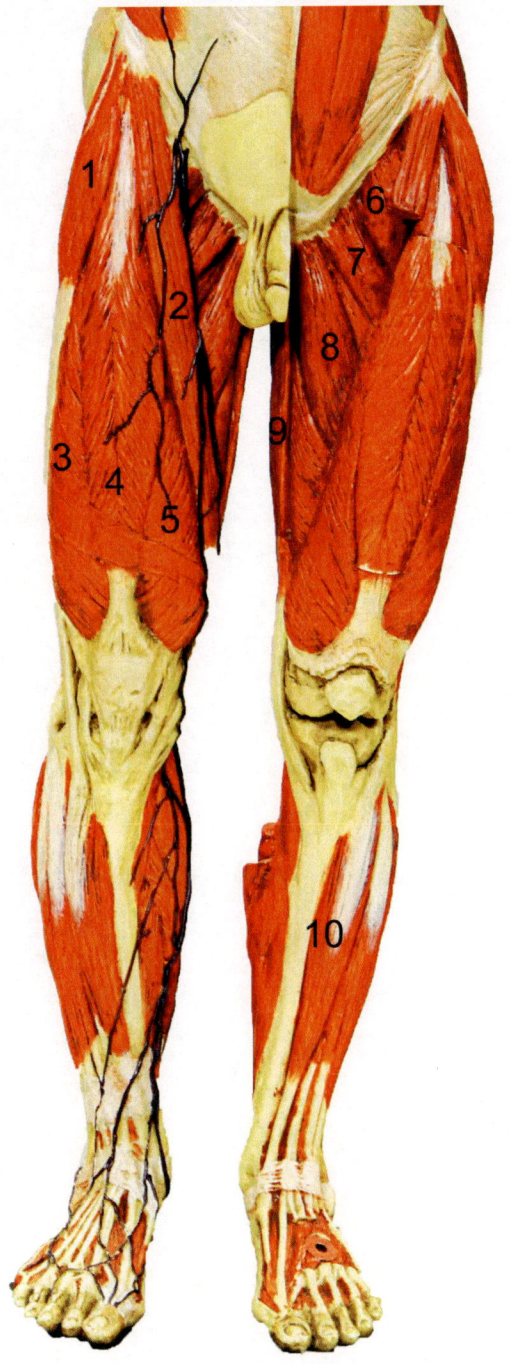

Figure 3-13. Lower Extremity - anterior
(Copyright 3B Scientific)

1. Tensor fasciae latae
2. Sartorius
3. Vastus lateralis
4. Rectus femoris
5. Vastus medialis
6. Iliopsoas
7. Pectineus
8. Adductor longus
9. Gracilis
10. Tibialis anterior

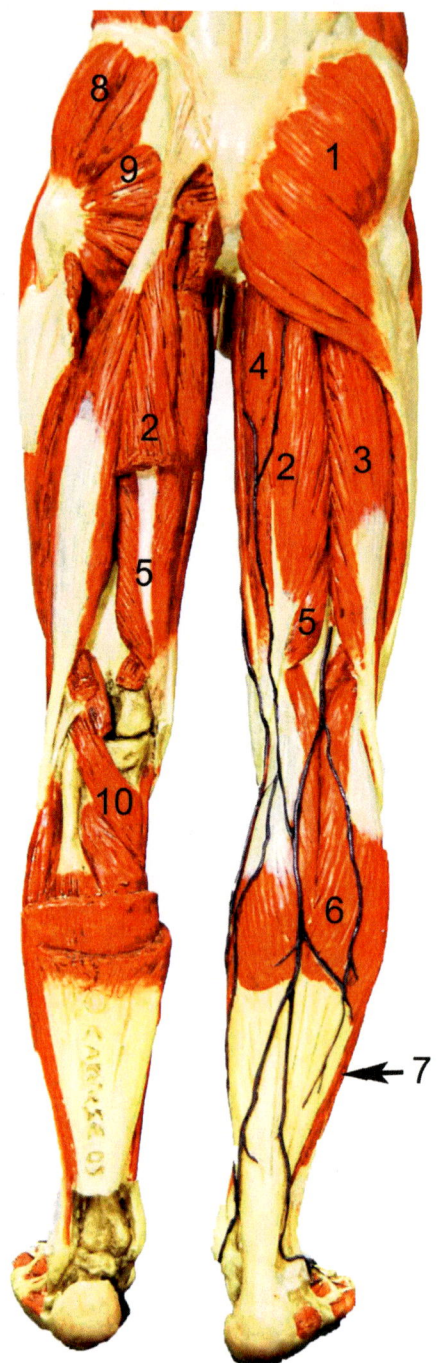

Figure 3-14. Lower Extremity - posterior
(Copyright 3B Scientific)

1. Gluteus maximus
2. Semitendinosus
3. Biceps femoris long head
4. Adductor magnus
5. Semimembranosus
6. Gastrocnemius
7. Soleus
8. Gluteus medius
9. Piriformis
10. Popliteus

Torso (3B)

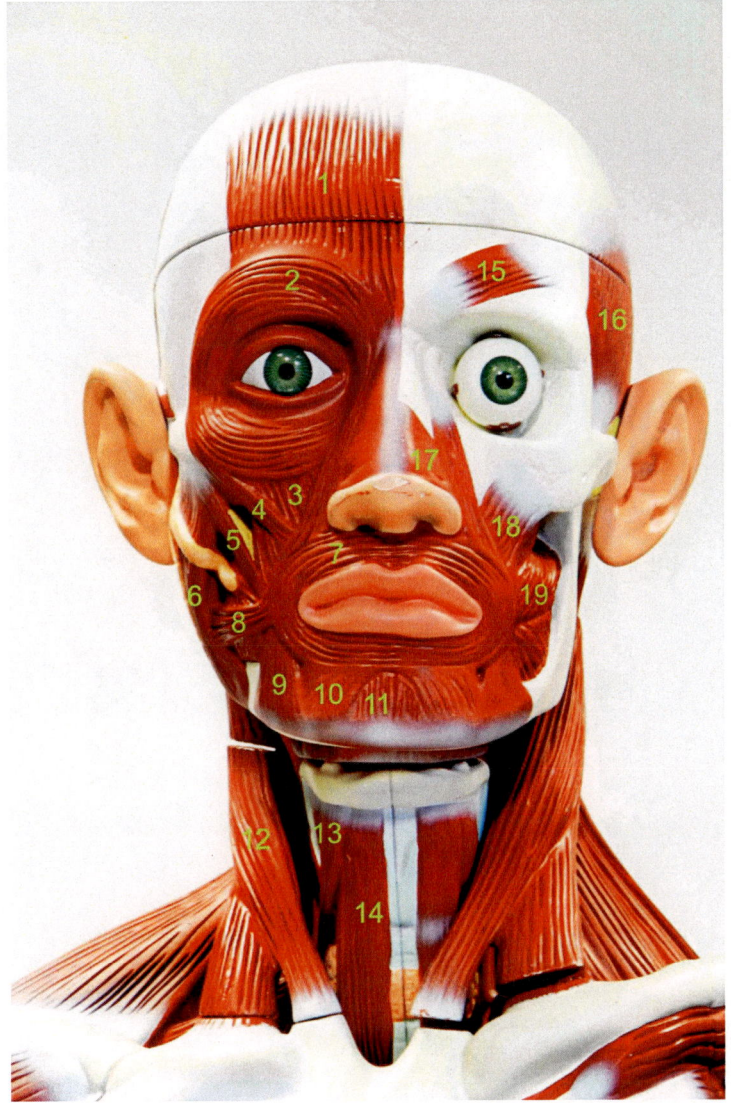

Figure 3-15. Head and Neck - anterior
(Copyright 3B Scientific)

1. Frontal belly of occipitofrontalis
2. Orbicularis oculi
3. Levator labii superioris
4. Zygomaticus minor
5. Zygomaticus major
6. Masseter
7. Orbicularis oris
8. Risorius
9. Depressor anguli oris
10. Depressor labii inferioris
11. Mentalis
12. Sternocleidomastoid
13. Omohyoid superior belly
14. Sternohyoid
15. Corrugator supercilii
16. Temporalis
17. Nasalis
18. Levator anguli oris
19. Buccinator

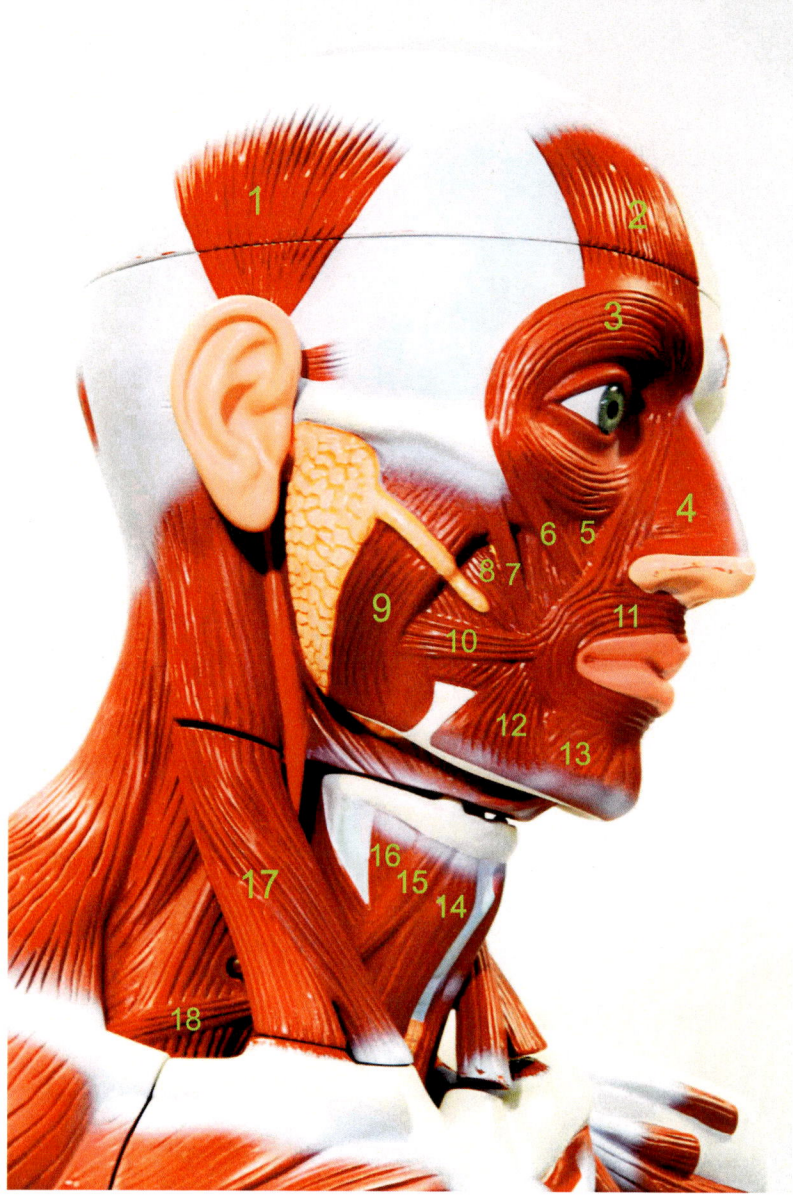

Figure 3-16. Head and Neck - lateral superficial muscles
(Copyright 3B Scientific)

1. Auricularis
2. Frontal belly of occipitofrontalis
3. Orbicularis oculi
4. Nasalis
5. Levator labii superioris
6. Zygomaticus minor
7. Zygomaticus major
8. Buccinator
9. Masseter
10. Risorius
11. Orbicularis oris
12. Depressor anguli oris
13. Depressor labii inferioris
14. Sternohyoid
15. Omohyoid superior belly
16. Thyrohyoid
17. Sternocleidomastoid
18. Omohyoid inferior belly

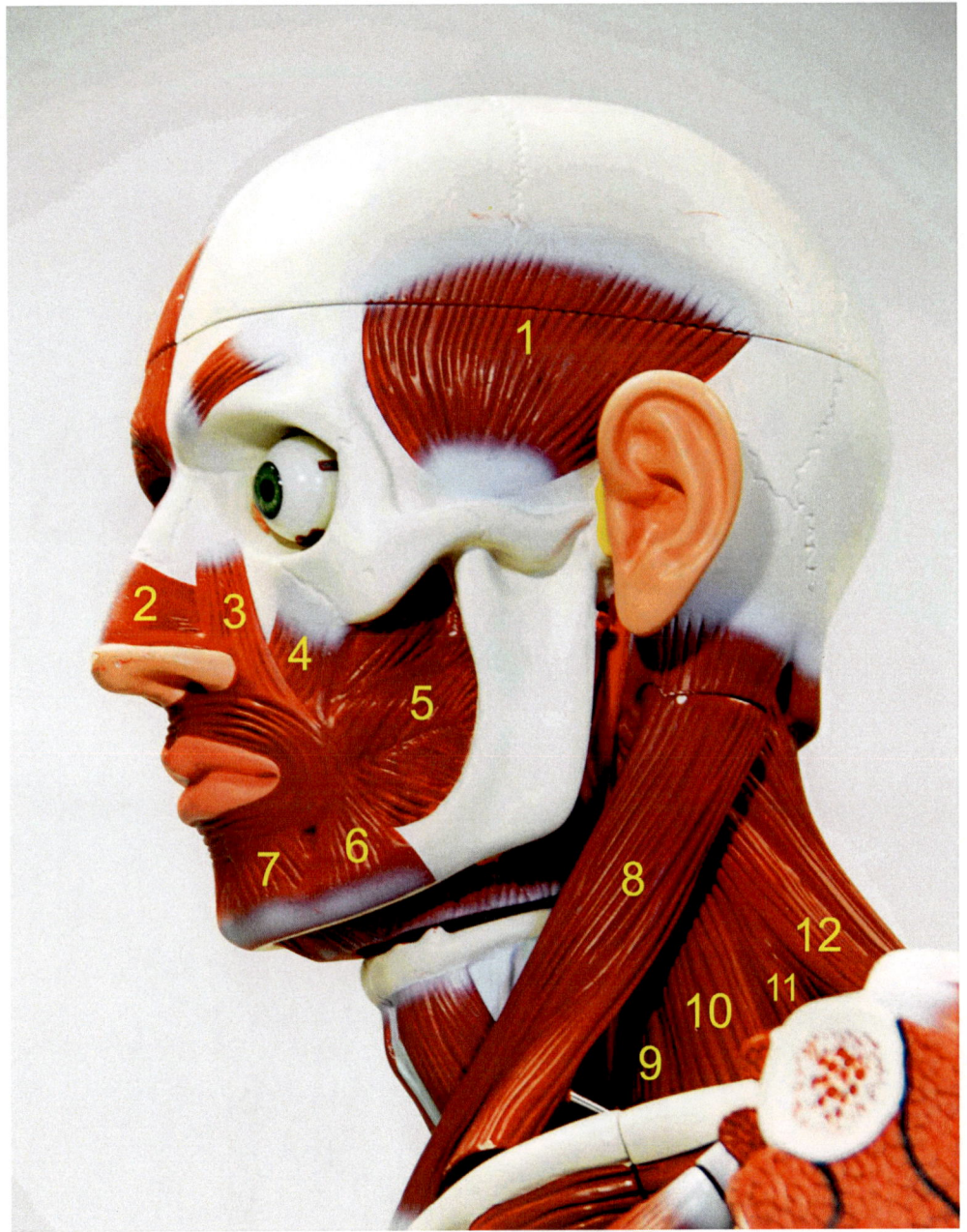

Figure 3-17. Head and Neck - lateral deep muscles
(Copyright 3B Scientific)

1. Temporalis
2. Nasalis
3. Levator labii superioris alaeque nasi
4. Levator anguli oris
5. Buccinator
6. Depressor anguli oris
7. Depressor labii inferioris
8. Sternocleidomastoid
9. Anterior scalene
10. Middle scalene
11. Posterior scalene
12. Levator scapulae

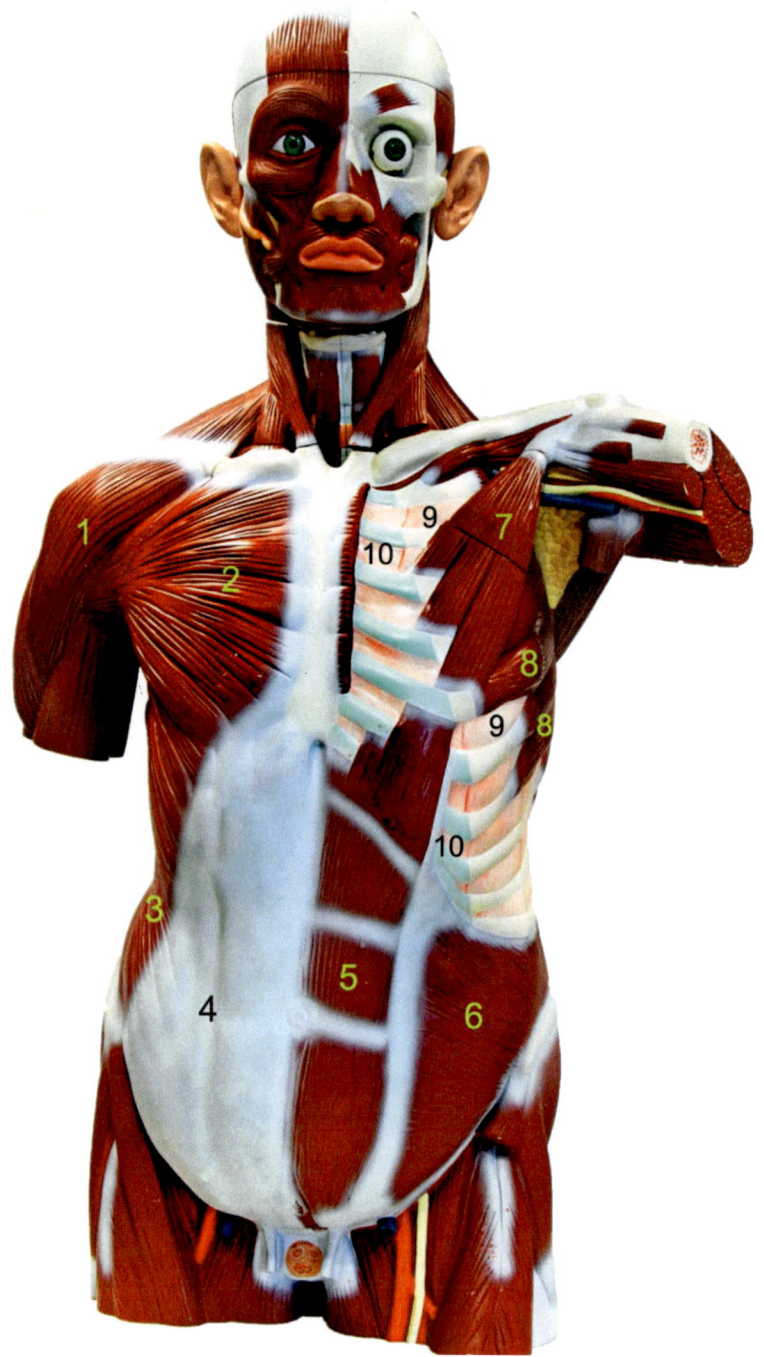

Figure 3-18. Torso – anterior
(Copyright 3B Scientific)

1. Deltoid
2. Pectoralis major
3. External oblique
4. Aponeurosis of external oblique
5. Rectus abdominis
6. Internal oblique
7. Pectoralis minor
8. Serratus anterior
9. External intercostal
10. Internal intercostal

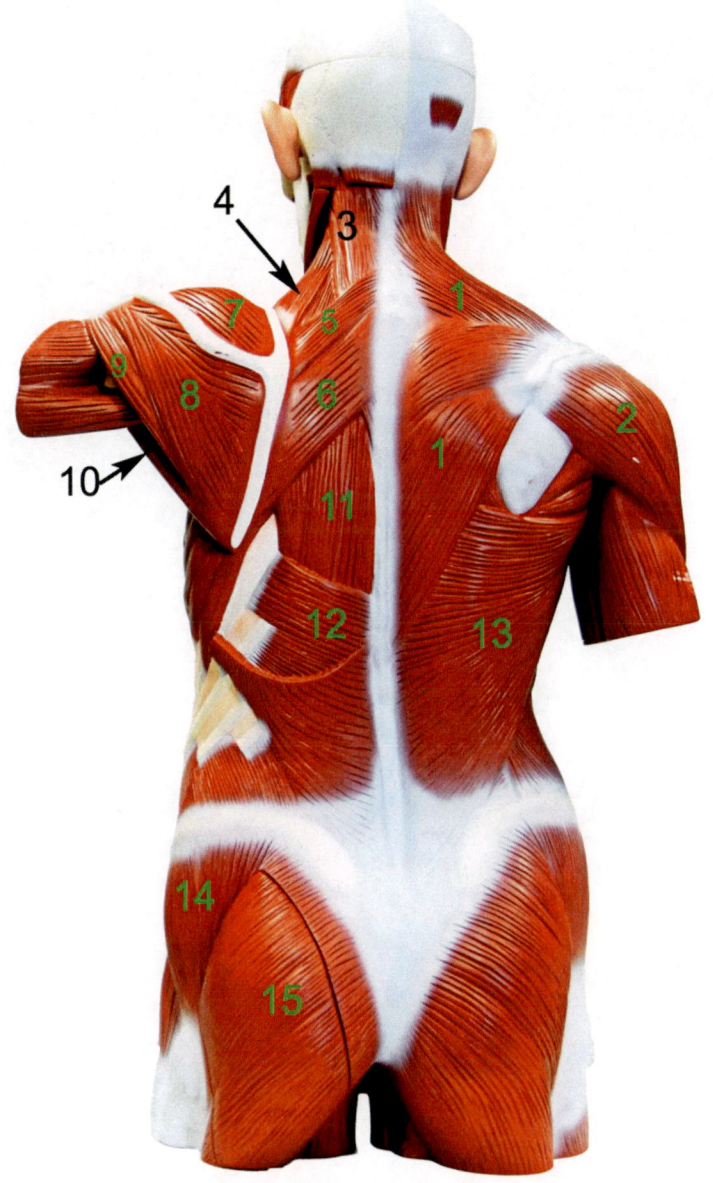

Figure 3-19. Torso - posterior
(Copyright 3B Scientific)

1. Trapezius
2. Deltoid
3. Splenius capitis
4. Levator scapulae
5. Rhomboid minor
6. Rhomboid major
7. Supraspinatus
8. Infraspinatus
9. Teres minor
10. Teres major
11. Erector spinae
12. Serratus posterior inferior
13. Latissimus dorsi
14. Gluteus medius
15. Gluteus maximus

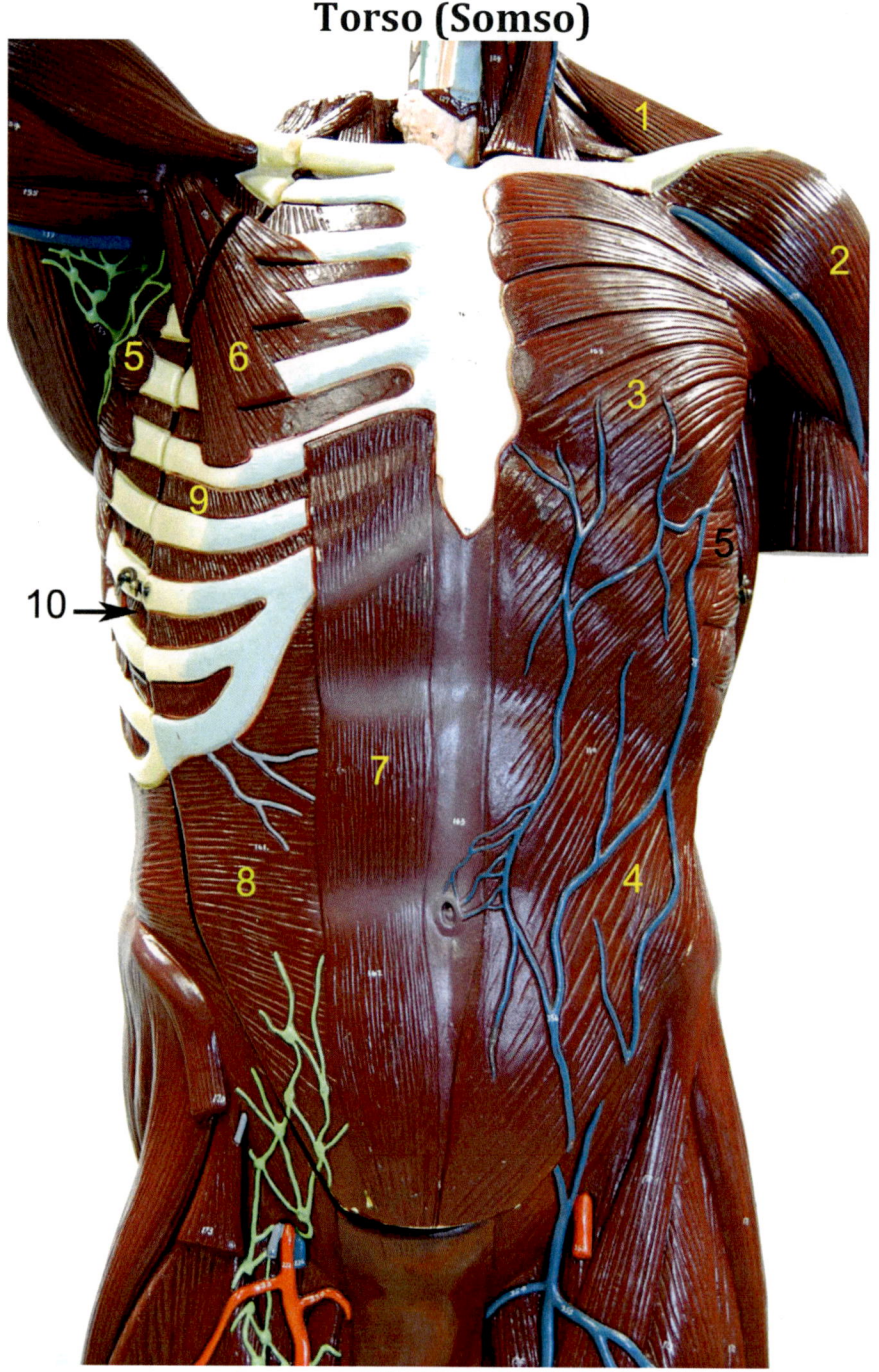

Figure 3-20. Trunk - anterior
(Copyright Somso)

1. Trapezius
2. Deltoid
3. Pectoralis major
4. External oblique
5. Serratus anterior
6. Pectoralis minor
7. Rectus abdominis
8. Internal oblique
9. External intercostal
10. Internal intercostal

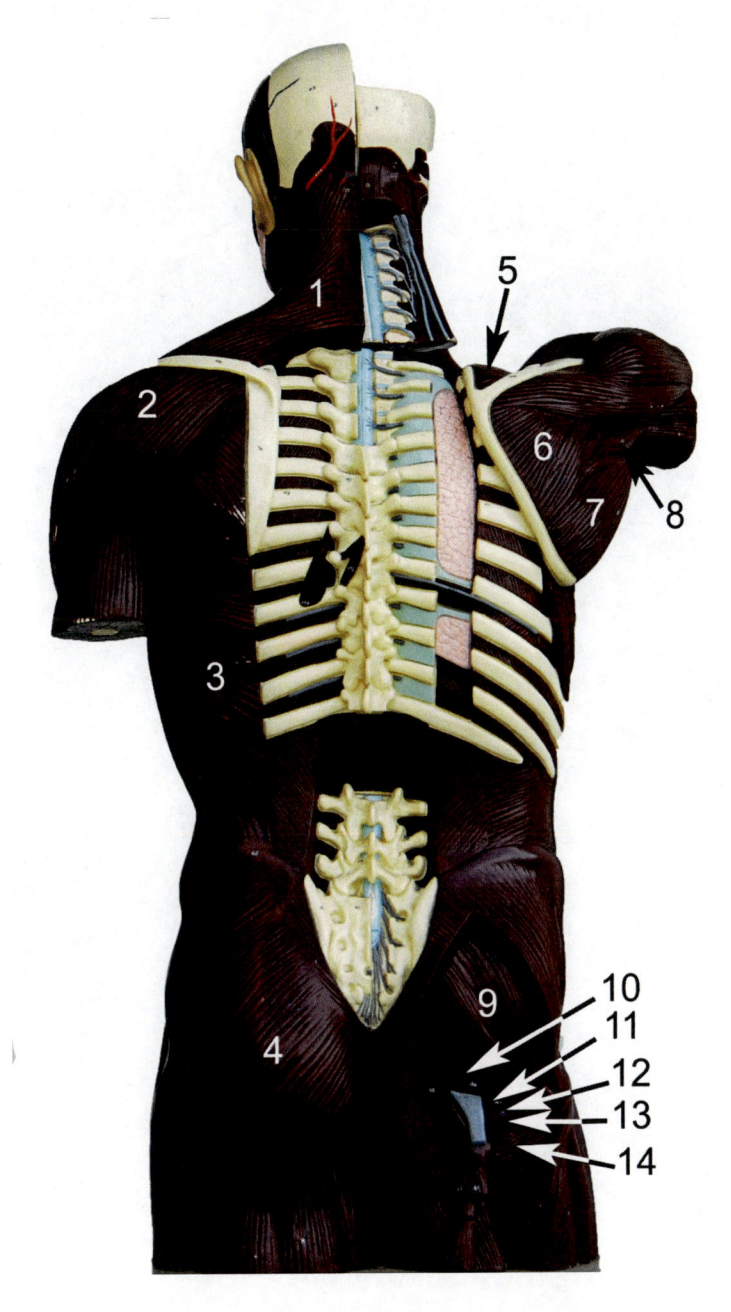

Figure 3-21. Trunk – posterior
(Copyright Somso)

1. Trapezius
2. Deltoid
3. Latissimus dorsi
4. Gluteus maximus
5. Supraspinatus
6. Infraspinatus
7. Teres major
8. Teres minor
9. Gluteus medius
10. Piriformis
11. Superior gemellus
12. Obturator internus
13. Inferior gemellus
14. Quadratus femoris

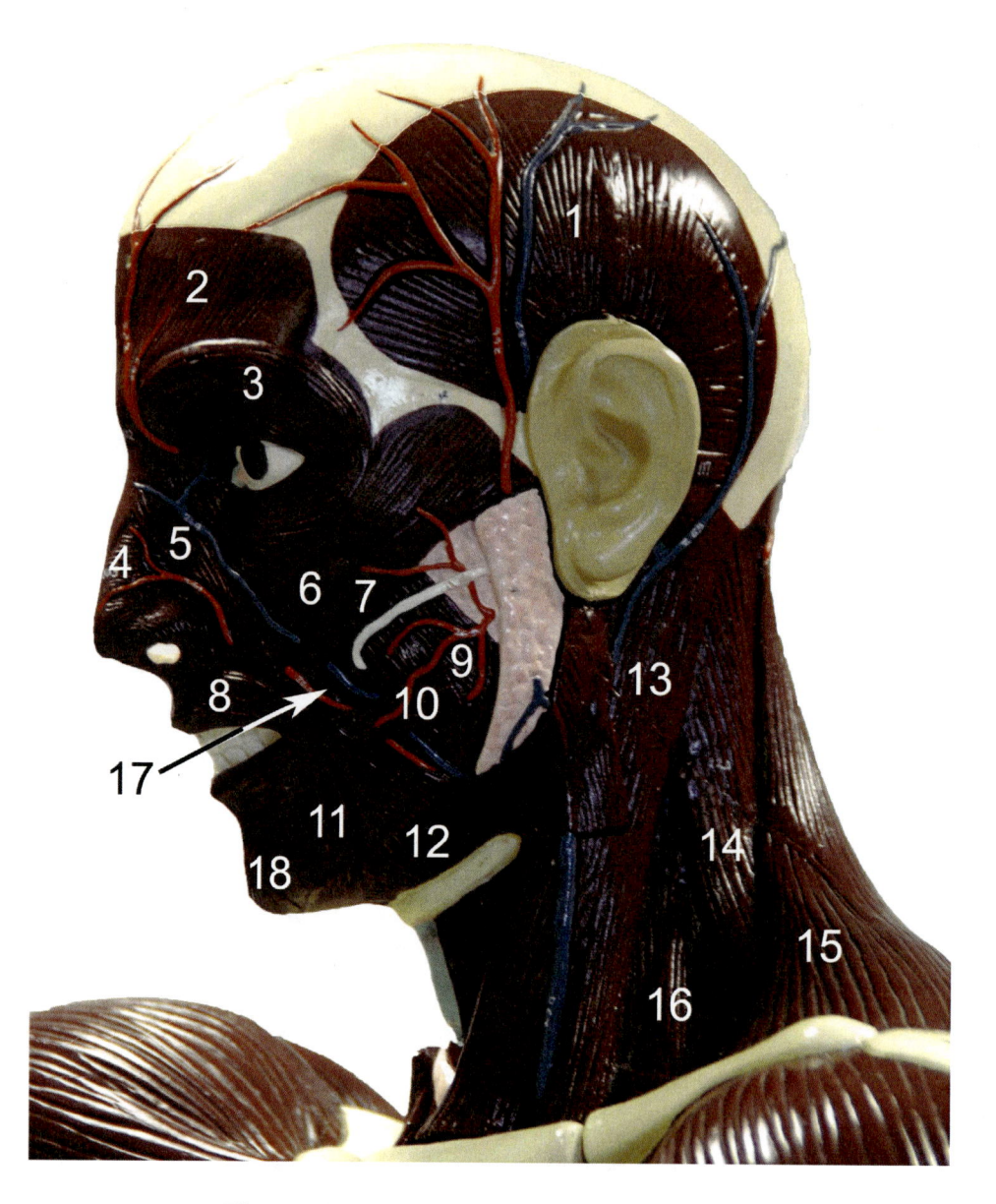

Figure 3-22. Head and Neck - lateral
(Copyright Somso)

1. Auricularis
2. Frontal belly of occipitofrontalis
3. Orbicularis oculi
4. Nasalis
5. Levator labii superioris alaeque nasi
6. Zygomaticus minor
7. Zygomaticus major
8. Orbicularis oris
9. Masseter
10. Risorius
11. Depressor anguli oris
12. Platysma (cut)
13. Sternocleidomastoid
14. Splenius capitis
15. Trapezius
16. Scalene muscles
17. Buccinator
18. Depressor labii inferioris

Torso (Denoyer-Geppert)

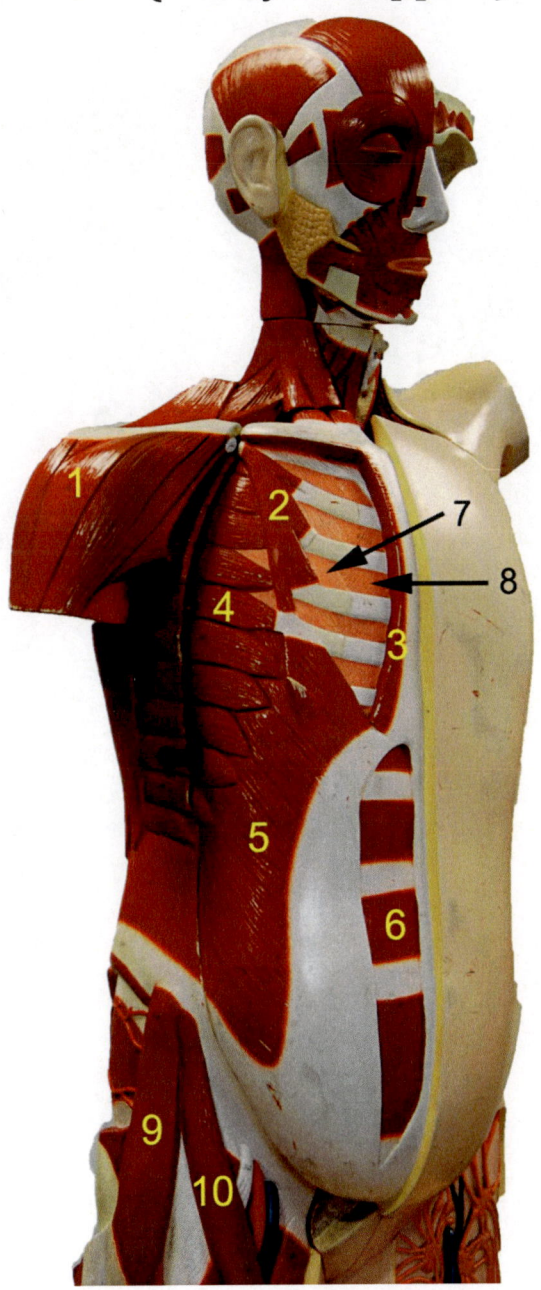

Figure 3-23. Trunk - anterior
(Copyright Denoyer Geppert)

1. Deltoid
2. Pectoralis minor
3. Pectoralis major (cut)
4. Serratus anterior
5. External oblique
6. Rectus abdominis
7. External intercostal
8. Internal intercostal
9. Tensor fasciae latae
10. Sartorius

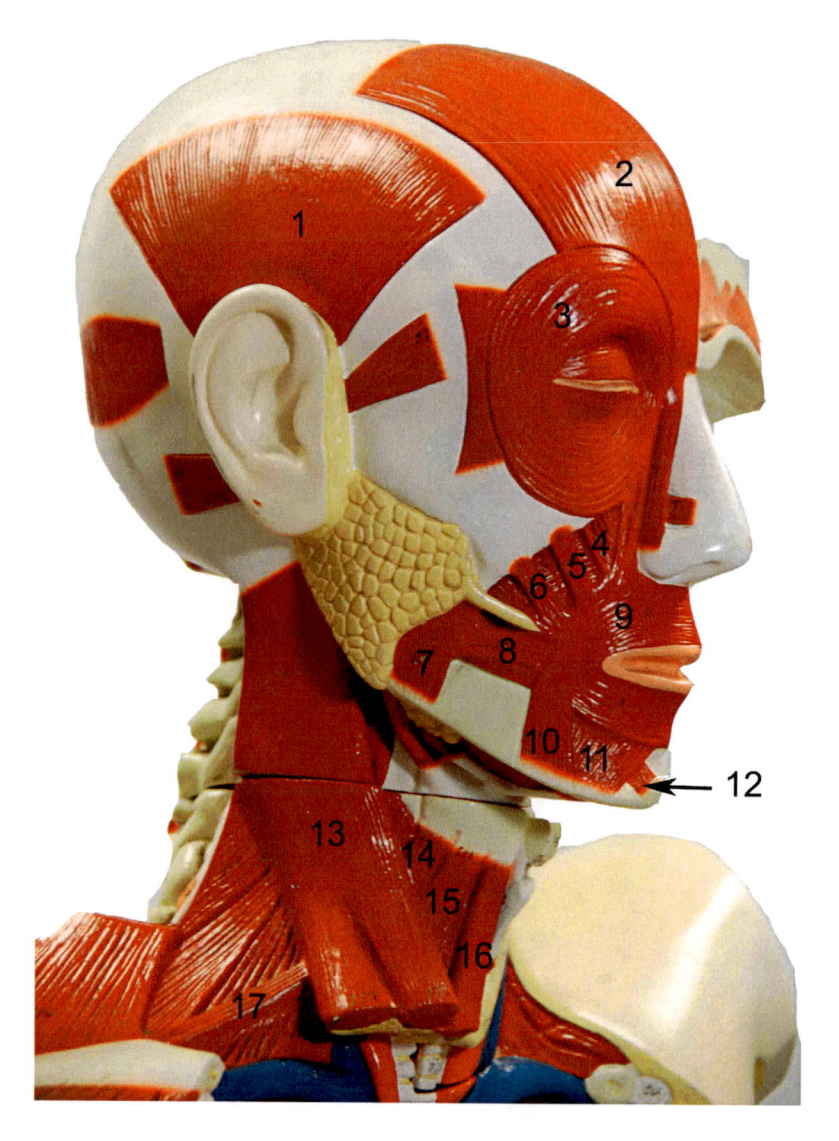

Figure 3-24. Head and Neck - lateral
(Copyright Denoyer Geppert)

1. Auricularis
2. Frontal belly of occipitofrontalis
3. Orbicularis oculi
4. Levator labii superioris
5. Zygomaticus minor
6. Zygomaticus major
7. Masseter
8. Risorius
9. Orbicularis oris
10. Depressor anguli oris
11. Depressor labii inferioris
12. Mentalis
13. Sternocleidomastoid
14. Thyrohyoid
15. Omohyoid superior belly
16. Sternohyoid
17. Omohyoid inferior belly

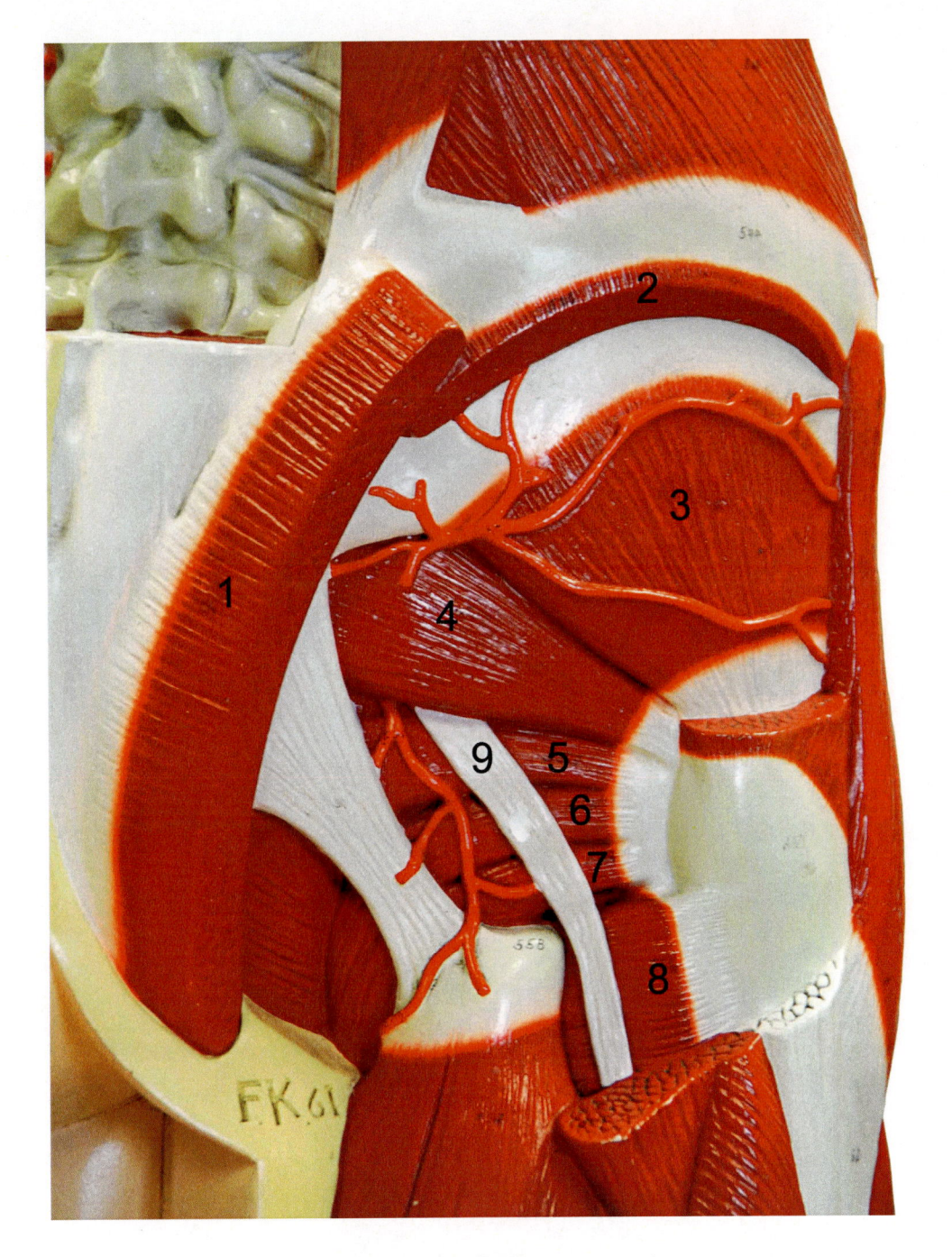

Figure 3-25. Hip - posterior
(Copyright Denoyer Geppert)

1. Gluteus maximus (cut)
2. Gluteus medius (cut)
3. Gluteus minimus
4. Piriformis
5. Superior gemellus
6. Obturator internus
7. Inferior gemellus
8. Quadratus femoris
9. Sciatic nerve

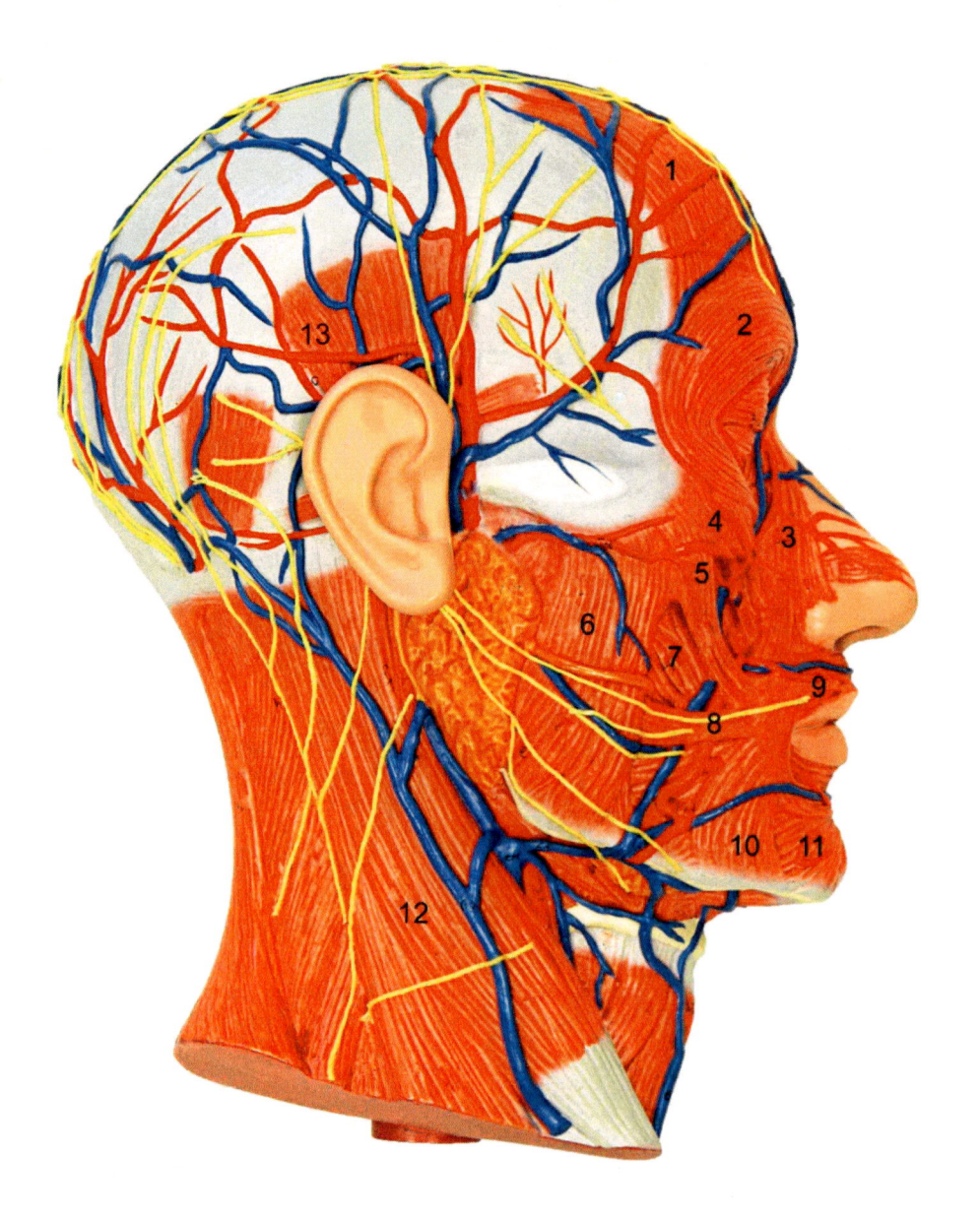

Figure 3-26. Half Head with Musculature
(Copyright 3B Scientific)

1. Frontal belly of occipitofrontalis
2. Orbicularis oculi
3. Levator labii superioris alaeque nasi
4. Zygomaticus minor
5. Zygomaticus major
6. Masseter
7. Buccinator
8. Risorius
9. Orbicularis oris
10. Depressor anguli oris
11. Depressor labii inferioris
12. Sternocleidomastoid
13. Auricularis

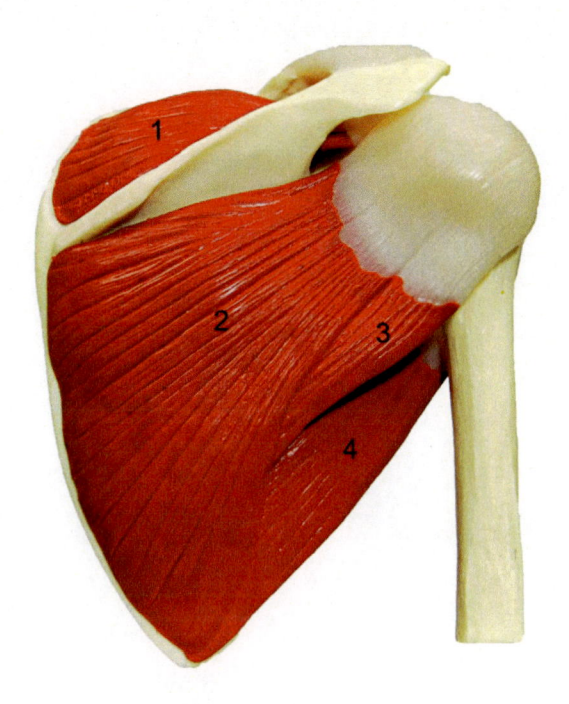

Figure 3-27. Scapula - posterior

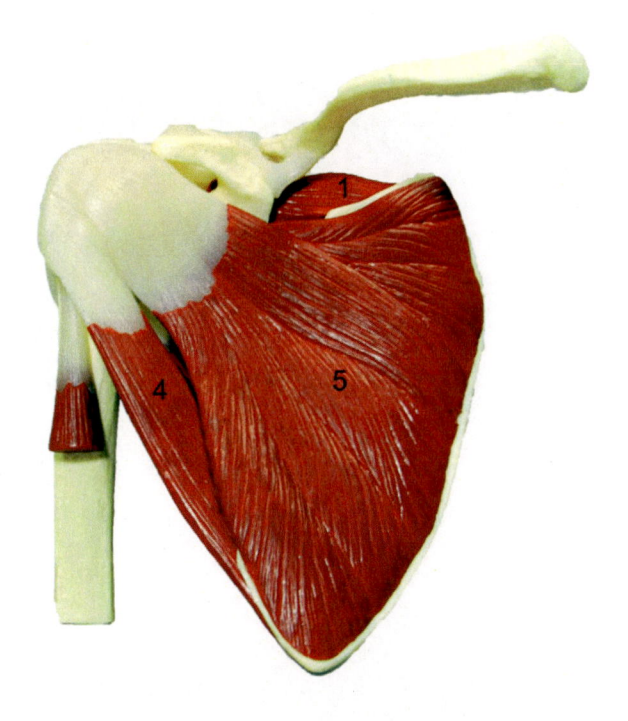

Figure 3-28. Scapula - anterior
(Copyright GFI Anatomicals)

1. Supraspinatus
2. Infraspinatus
3. Teres minor
4. Teres major
5. Subscapularis

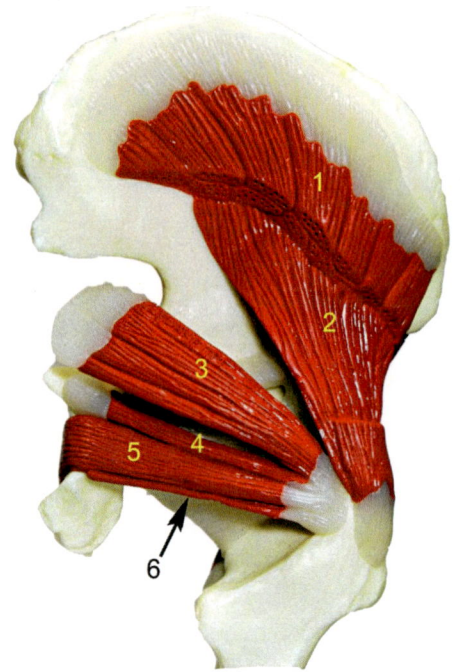

Figure 3-29. Hip - posterior

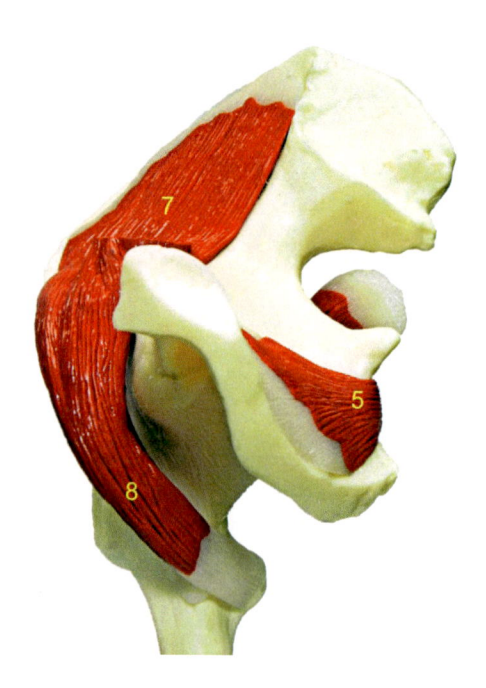

Figure 3-30. Hip – anterior
(Copyright GFI Anatomicals)

1. Gluteus medius (cut)
2. Gluteus minimus
3. Piriformis
4. Superior gemellus
5. Obturator internus
6. Inferior gemellus
7. Iliacus
8. Iliopsoas

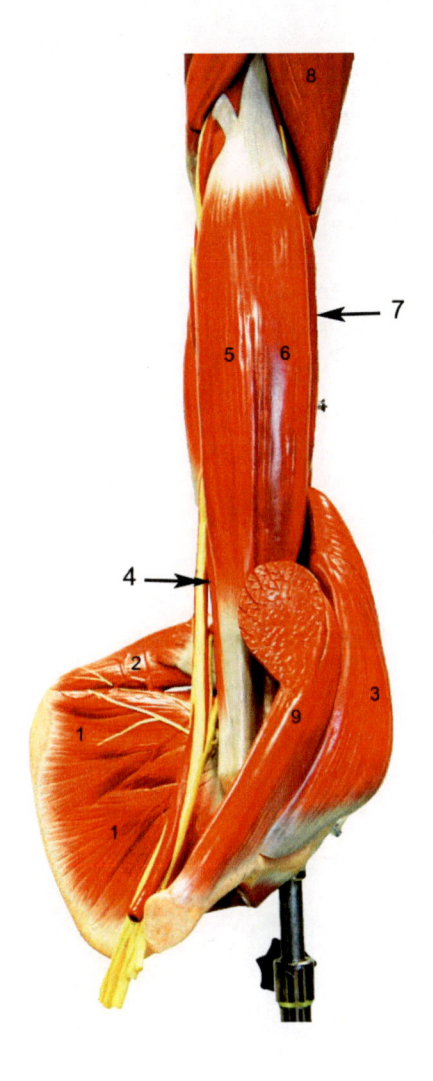

Figure 3-31. Arm - anterior
(Copyright Somso)

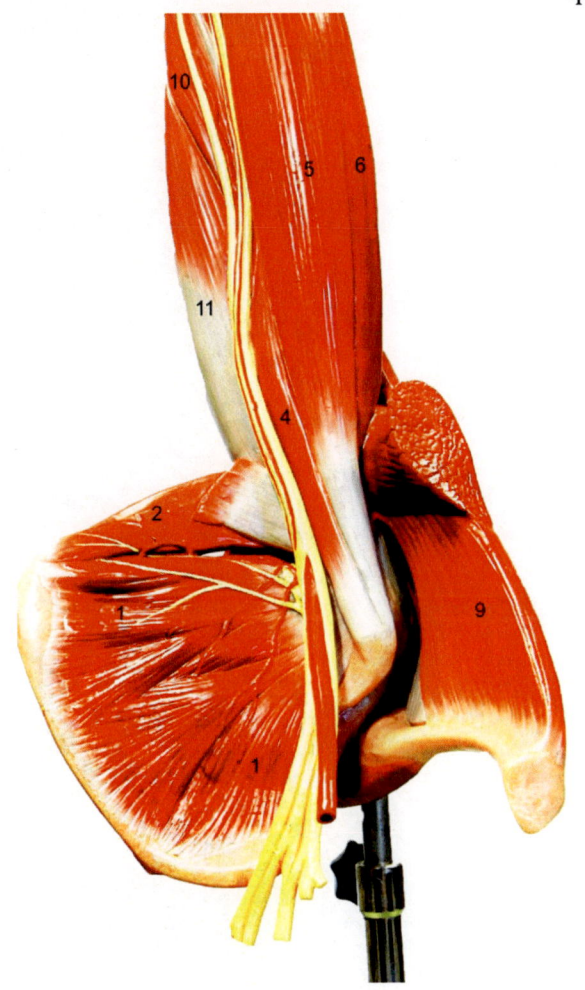

Figure 3-32. Arm - medial

1. Subscapularis
2. Teres major
3. Deltoid
4. Coracobrachialis
5. Biceps brachii short head
6. Biceps brachii long head
7. Brachialis
8. Brachioradialis
9. Pectoralis major (cut)
10. Triceps brachii medial head
11. Triceps brachii long head

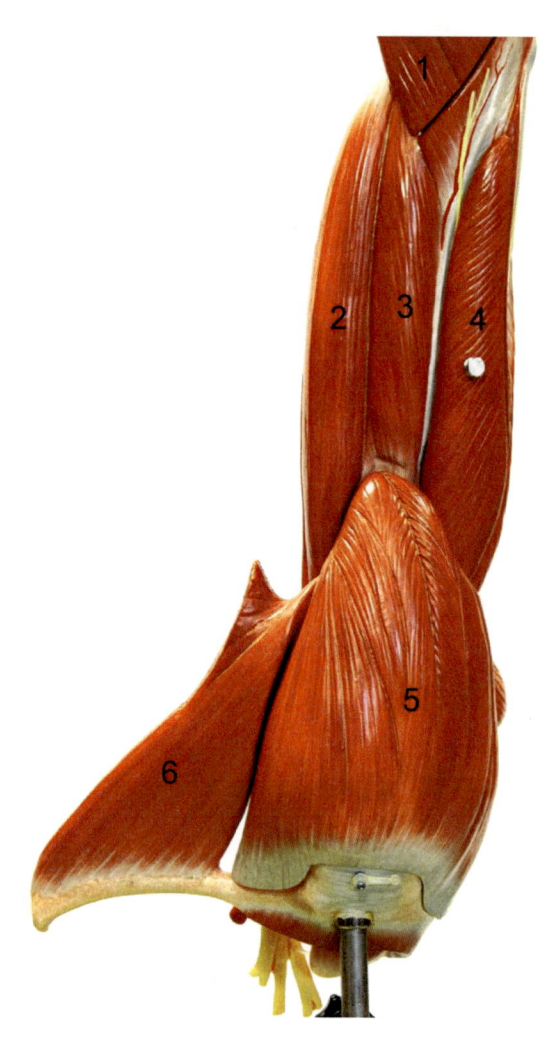

Figure 3-33. Arm - lateral
(Copyright Somso)

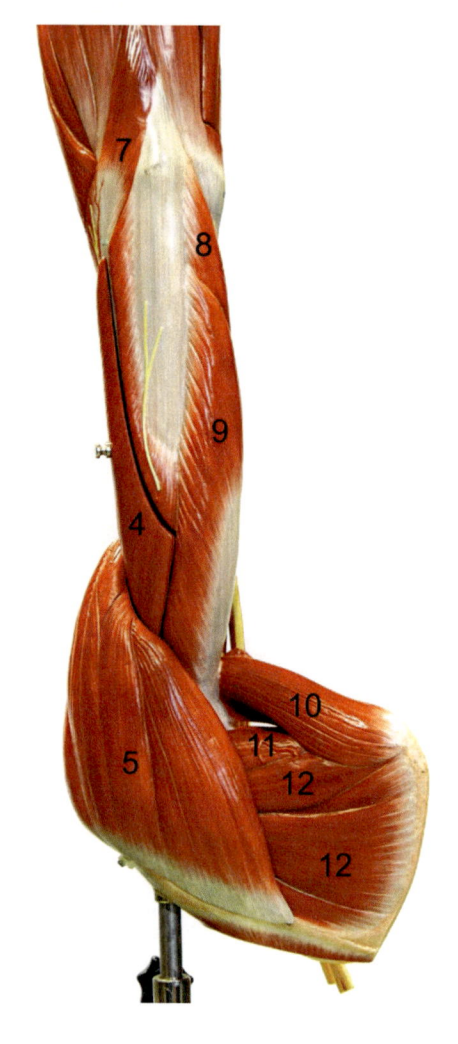

Figure 3-34. Arm - posterior

1. Brachioradialis
2. Biceps brachii long head
3. Brachialis
4. Triceps brachii lateral head
5. Deltoid
6. Pectoralis (cut)
7. Anconeus
8. Triceps brachii medial head
9. Triceps brachii long head
10. Teres major
11. Teres minor
12. Infraspinatus

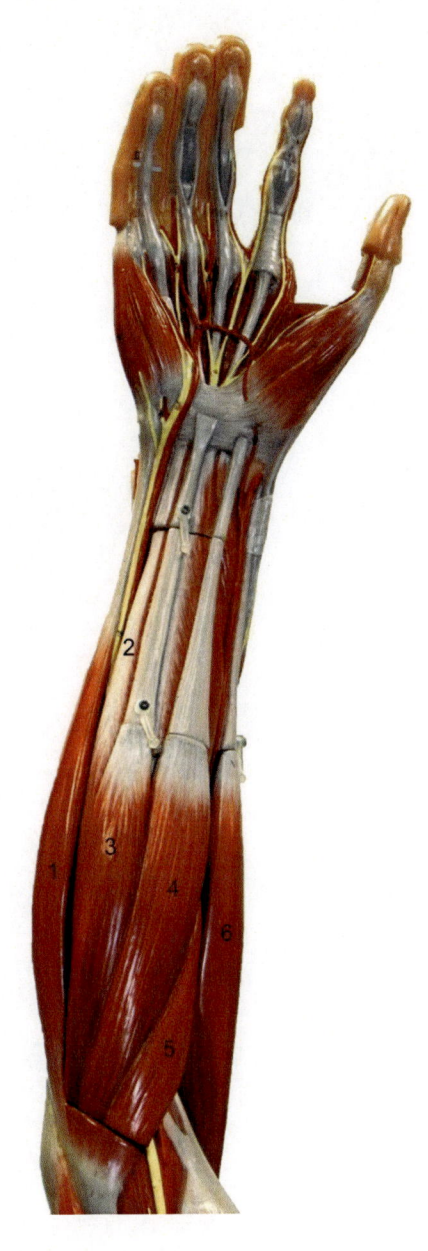

Figure 3-35. Forearm anterior
(Copyright Somso)

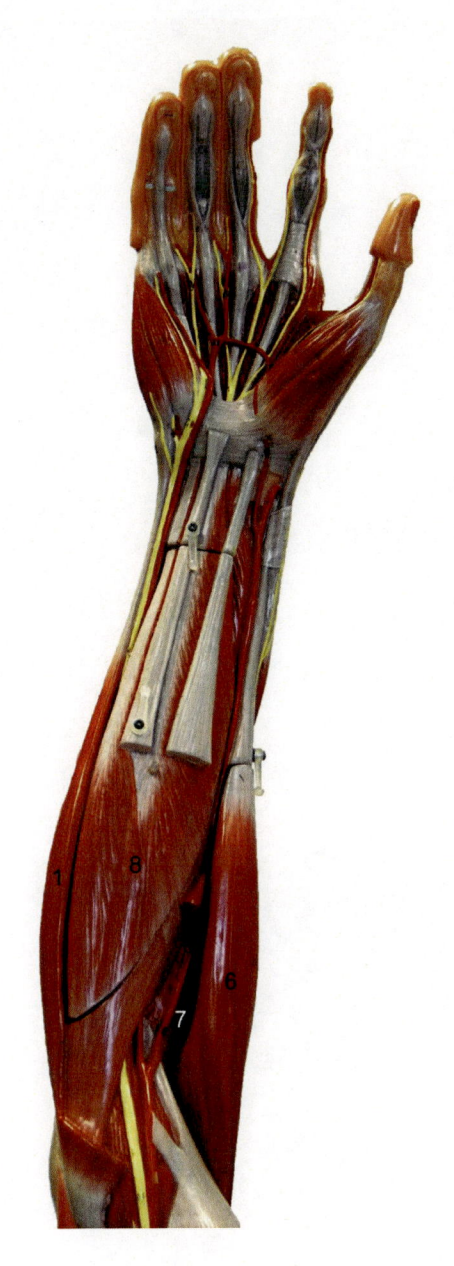

Figure 3-36. Forearm - anterior deep

1. Flexor carpi ulnaris
2. Tendon of flexor digitorum superficialis
3. Palmaris longus
4. Flexor carpi radialis
5. Pronator teres
6. Brachioradialis
7. Supinator
8. Flexor digitorum superficialis

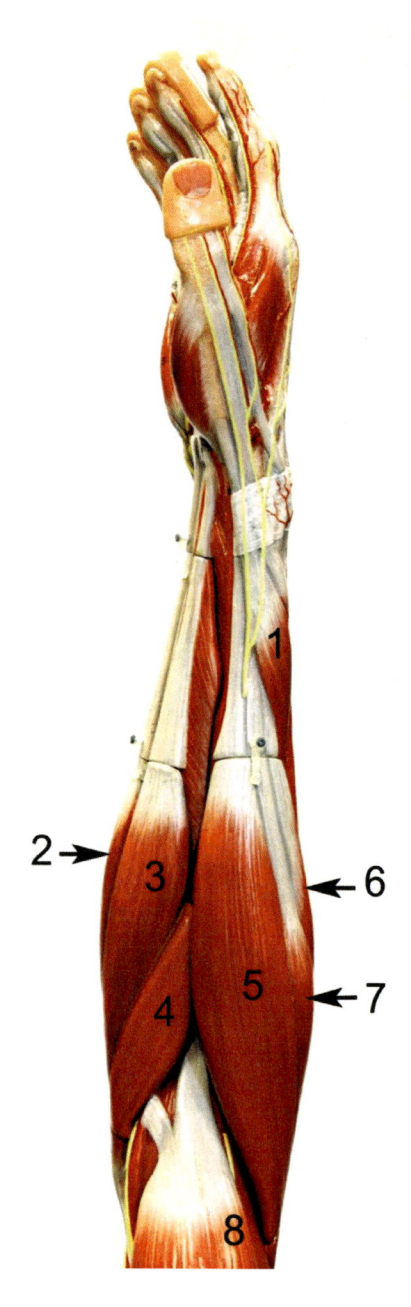

Figure 3-37. Forearm – lateral
(Copyright Somso)

1. Abductor pollicis longus
2. Palmaris longus
3. Flexor carpi radialis
4. Pronator teres
5. Brachioradialis
6. Extensor carpi radialis brevis

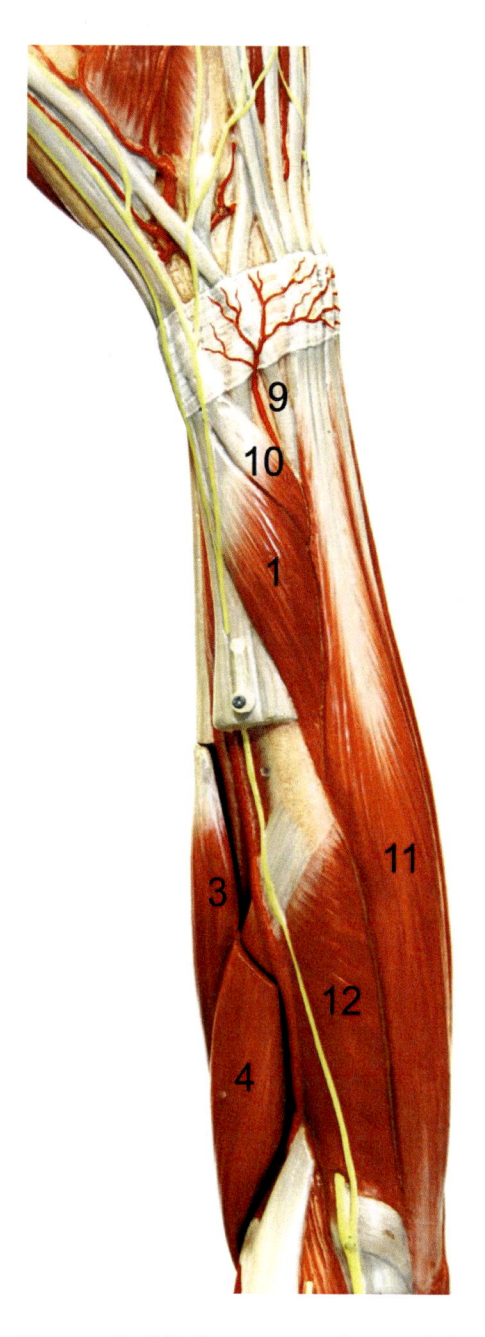

Figure 3-38. Forearm - lateral deep

7. Extensor carpi radialis longus
8. Biceps brachii long head
9. Tendon of extensor pollicis longus
10. Extensor pollicis brevis
11. Extensor digitorum
12. Supinator

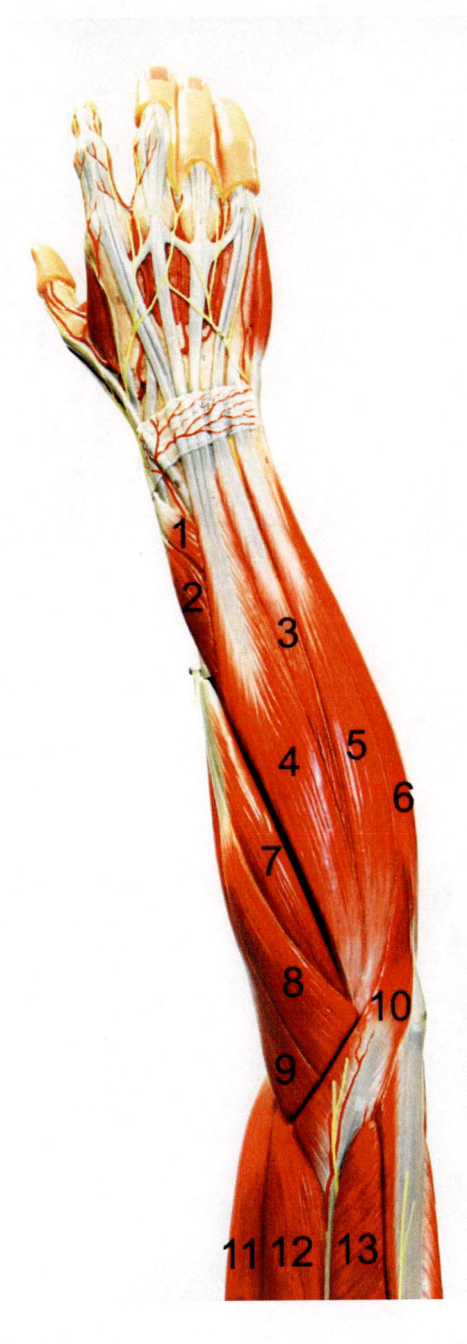

Figure 3-39. Forearm - posterior
(Copyright Somso)

1. Extensor pollicis brevis
2. Abductor pollicis longus
3. Extensor digiti minimi
4. Extensor digitorum
5. Extensor carpi ulnaris
6. Flexor carpi ulnaris
7. Extensor carpi radialis brevis
8. Extensor carpi radialis longus
9. Brachioradialis
10. Anconeus
11. Biceps brachii long head
12. Brachialis
13. Triceps brachii lateral head

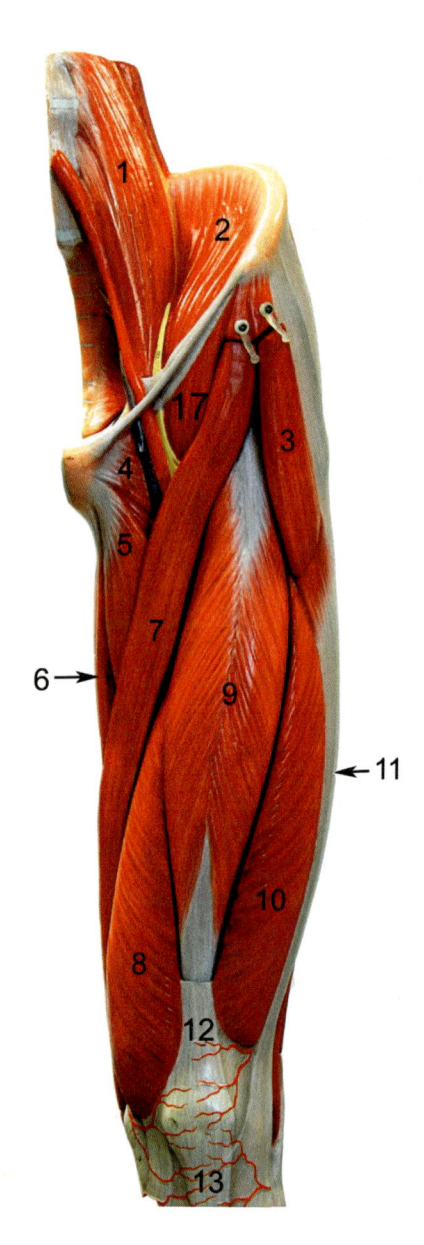

Figure 3-40. Thigh - anterior View

1. Psoas major
2. Iliacus
3. Tensor fasciae latae
4. Pectineus
5. Adductor longus
6. Gracilis
7. Sartorius
8. Vastus medialis
9. Rectus femoris
   (vastus intermedius is deep)

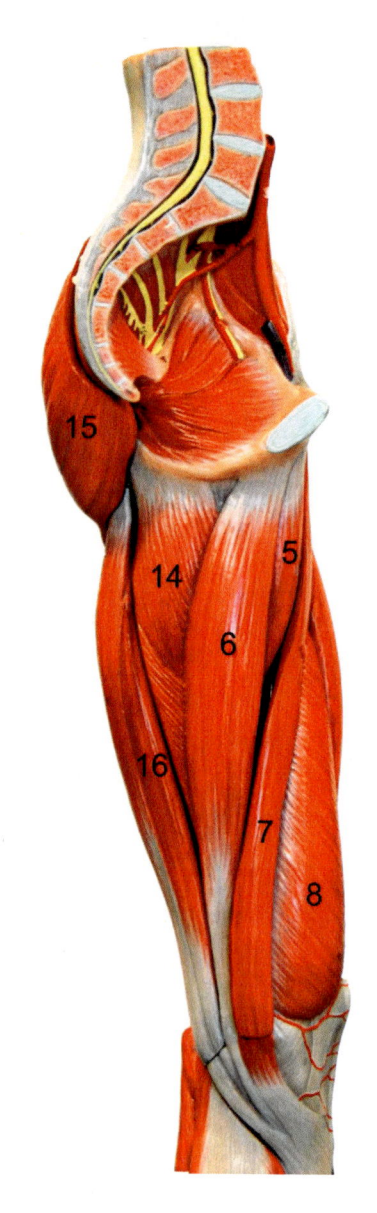

Figure 3-41. Thigh - medial
(Copyright Somso)

10. Vastus lateralis
11. Iliotibial tract (band)
12. Quadriceps tendon
13. Patellar ligament
14. Adductor magnus
15. Gluteus maximus
16. Semimembranosus
17. Iliopsoas

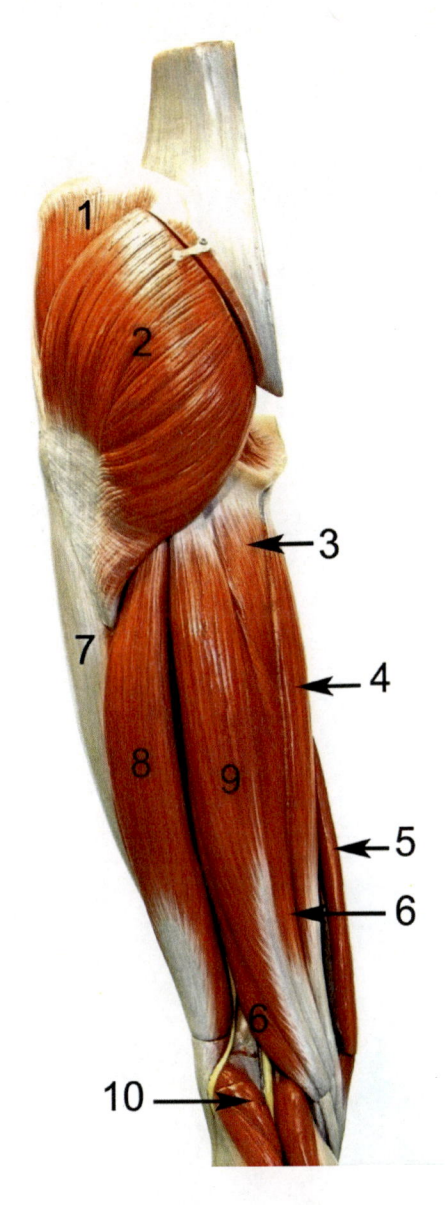

Figure 3-42. Thigh - posterior

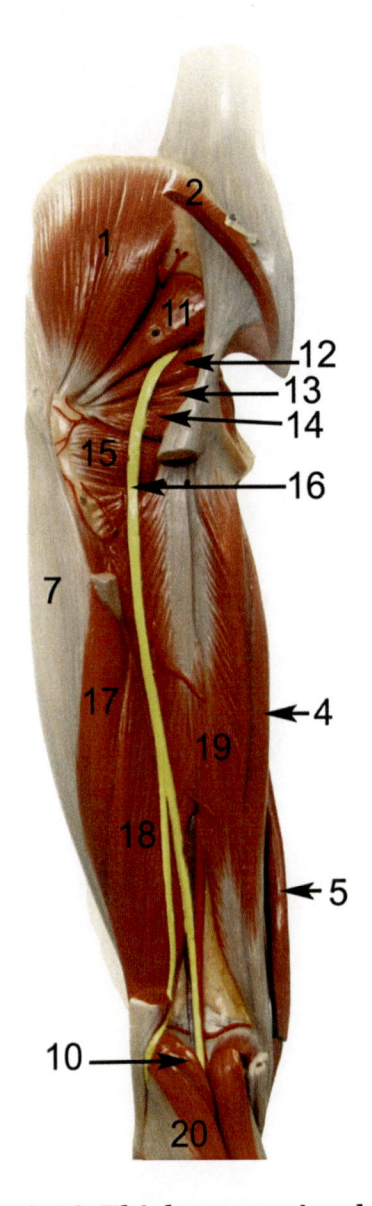

Figure 3-43. Thigh - posterior deep
(Copyright Somso)

1. Gluteus medius
2. Gluteus maximus
3. Adductor magnus
4. Gracilis
5. Sartorius
6. Semimembranosus
7. Iliotibial tract (band)
8. Biceps femoris long head
9. Semitendinosus
10. Plantaris
11. Piriformis
12. Superior gemellus
13. Obturator internus
14. Inferior gemellus
15. Quadratus femoris
16. Sciatic nerve
17. Vastus lateralis
18. Biceps femoris short head
19. Adductor magnus (deep)
20. Gastrocnemius

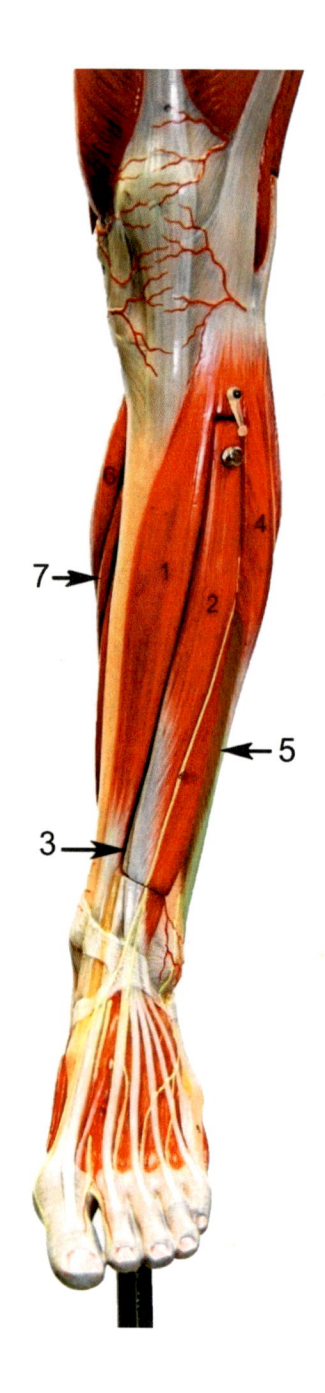

Figure 3-44. Leg - anterior

1. Tibialis anterior
2. Extensor digitorum longus
3. Extensor hallucis longus (deep to #2)

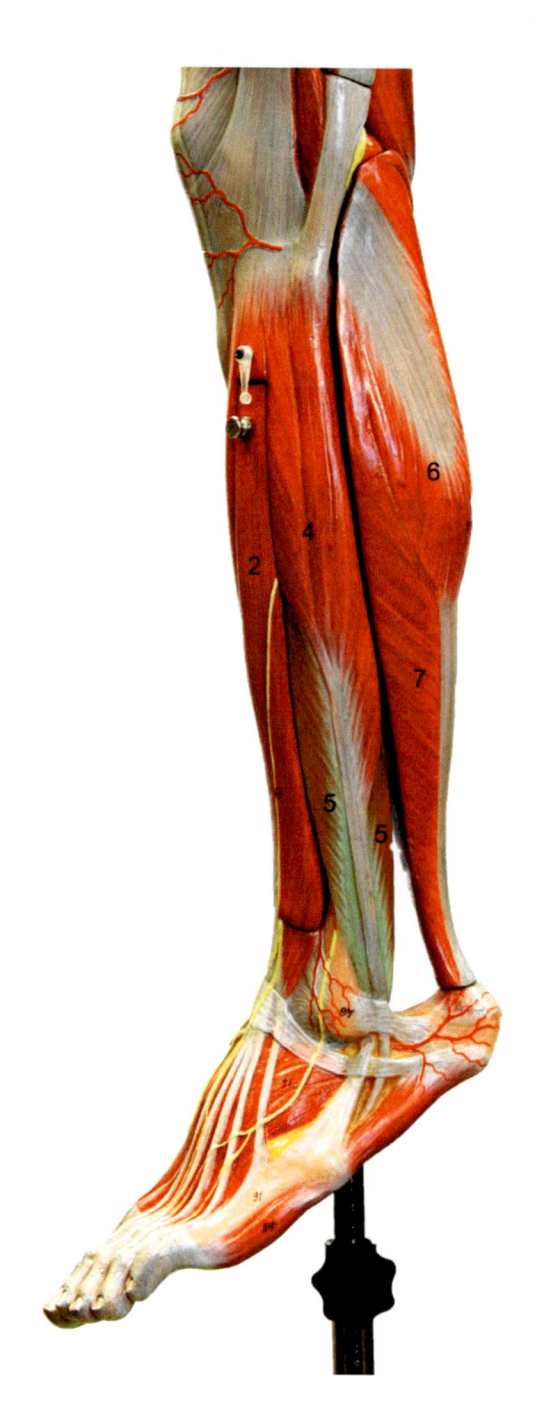

Figure 3-45. Leg - lateral
(Copyright Somso)

4. Fibularis (peroneus) longus
5. Fibularis (peroneus) brevis
6. Gastrocnemius
7. Soleus

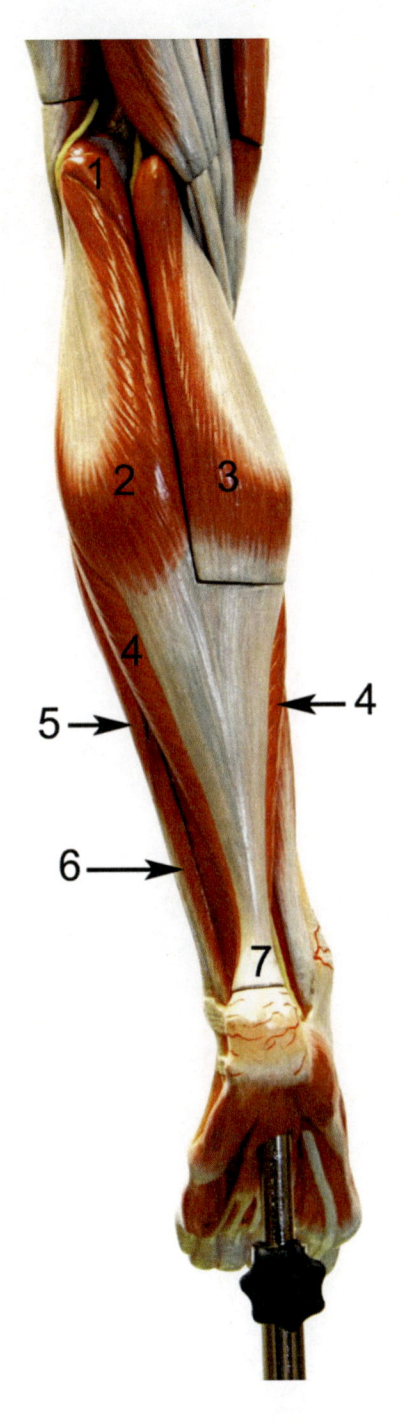

Figure 3-46. Leg - posterior
(Copyright Somso)
1. Plantaris
2. Gastrocnemius, lateral head
3. Gastrocnemius, medial head
4. Soleus
5. Fibularis (peroneus) longus
6. Fibularis (peroneus) brevis

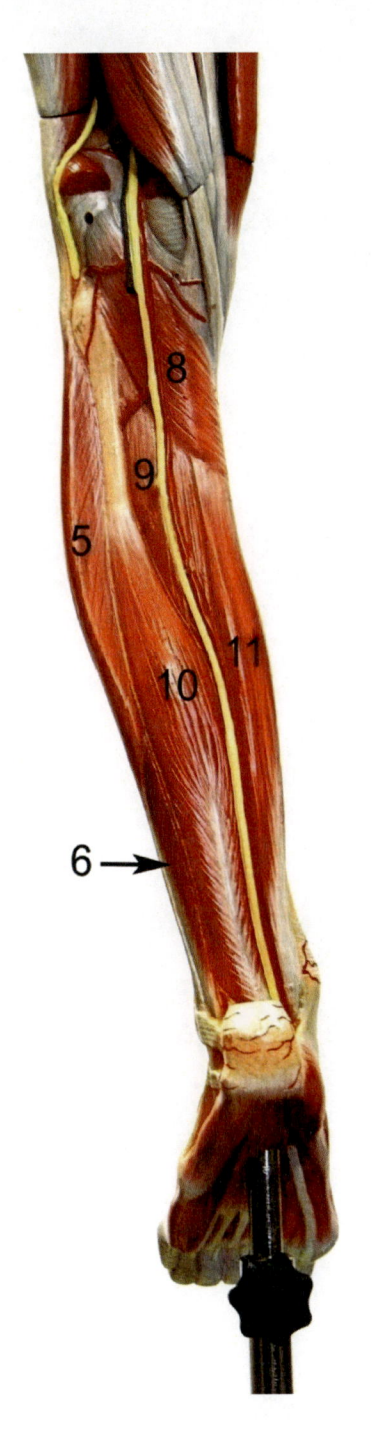

Figure 3-47. Leg - posterior deep

7. Calcaneal (Achilles) tendon
8. Popliteus
9. Tibialis posterior
10. Flexor hallucis longus
11. Flexor digitorum longus

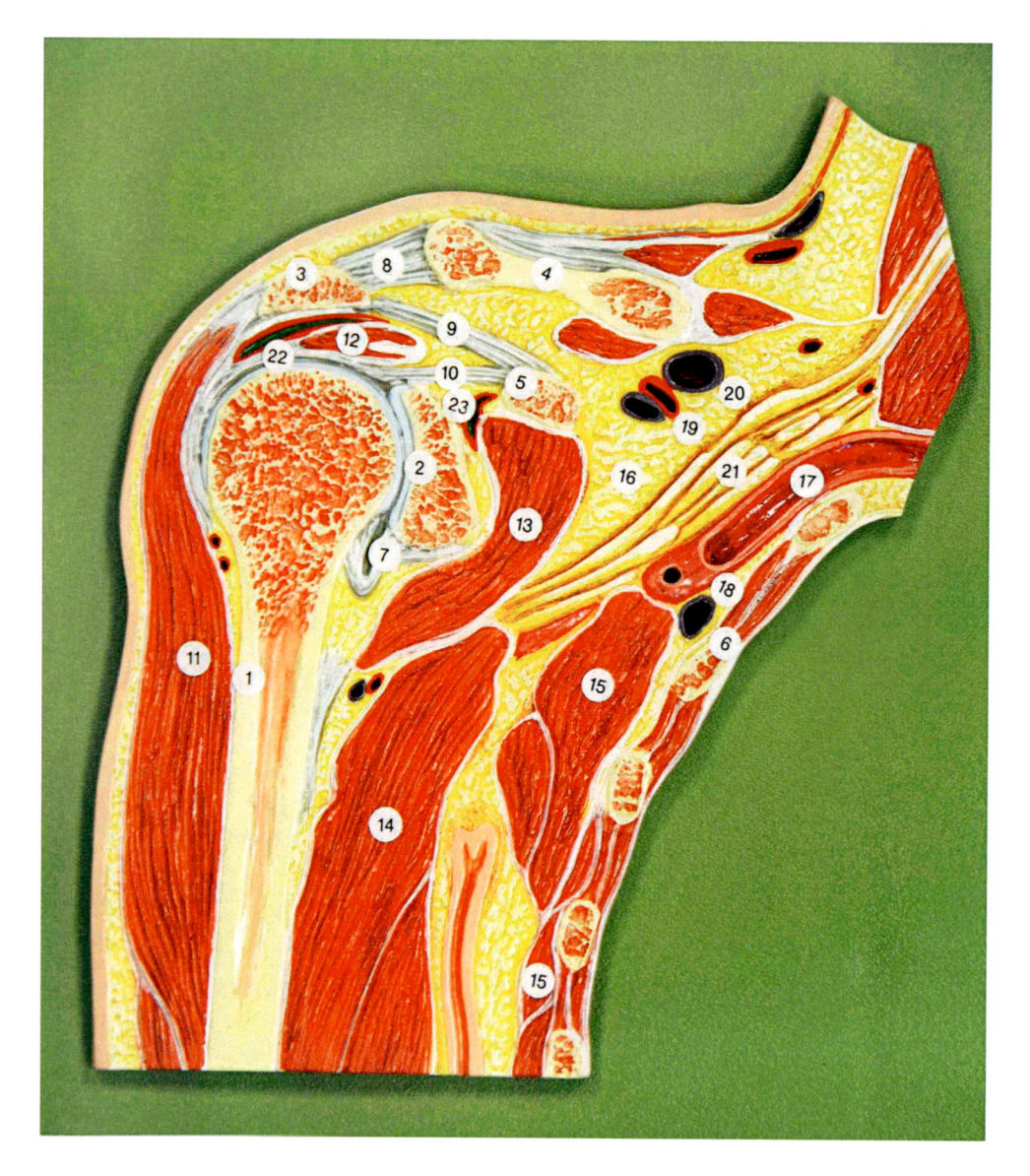

Figure 3-48. Shoulder Section
(Copyright Somso)

1. Humerus
2. Glenoid cavity
3. Acromion
4. Clavicle
5. Coracoid process
6. Ribs
7. Articular capsule
8. Acromioclavicular ligament
9. Coracoacromial ligament
10. Coracohumeral ligament
11. Deltoid
12. Supraspinatus
13. Subscapularis
14. Triceps brachii
15. Serratus anterior
16. Axillary cavity
17. Axillary artery
18. Subclavian vein
19. Suprascapular artery
20. Suprascapular nerve
21. Brachial plexus
22. Subdeltoid bursa
23. Subcoracoid bursa

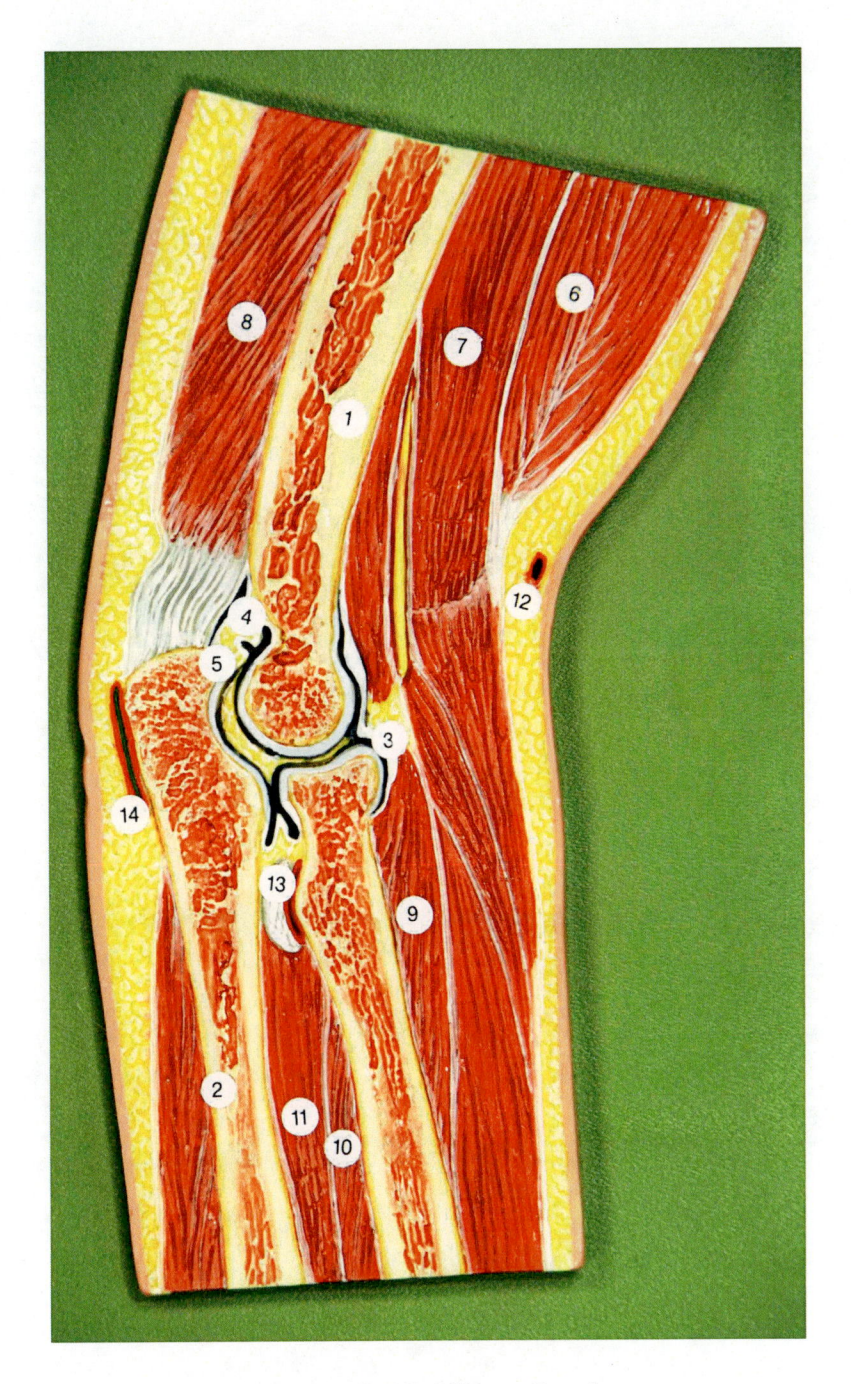

Figure 3-49. Elbow Section
(Copyright Somso)

1. Humerus
2. Ulna
3. Anterior capsular wall
4. Posterior capsular wall
5. Fatty fold of scapular wall
6. Biceps brachii
7. Brachialis
8. Triceps brachii
9. Pronator teres
10. Flexor digitorum superficialis
11. Flexor digitorum profundus
12. Radial artery
13. Ulnar artery
14. Synovial bursa

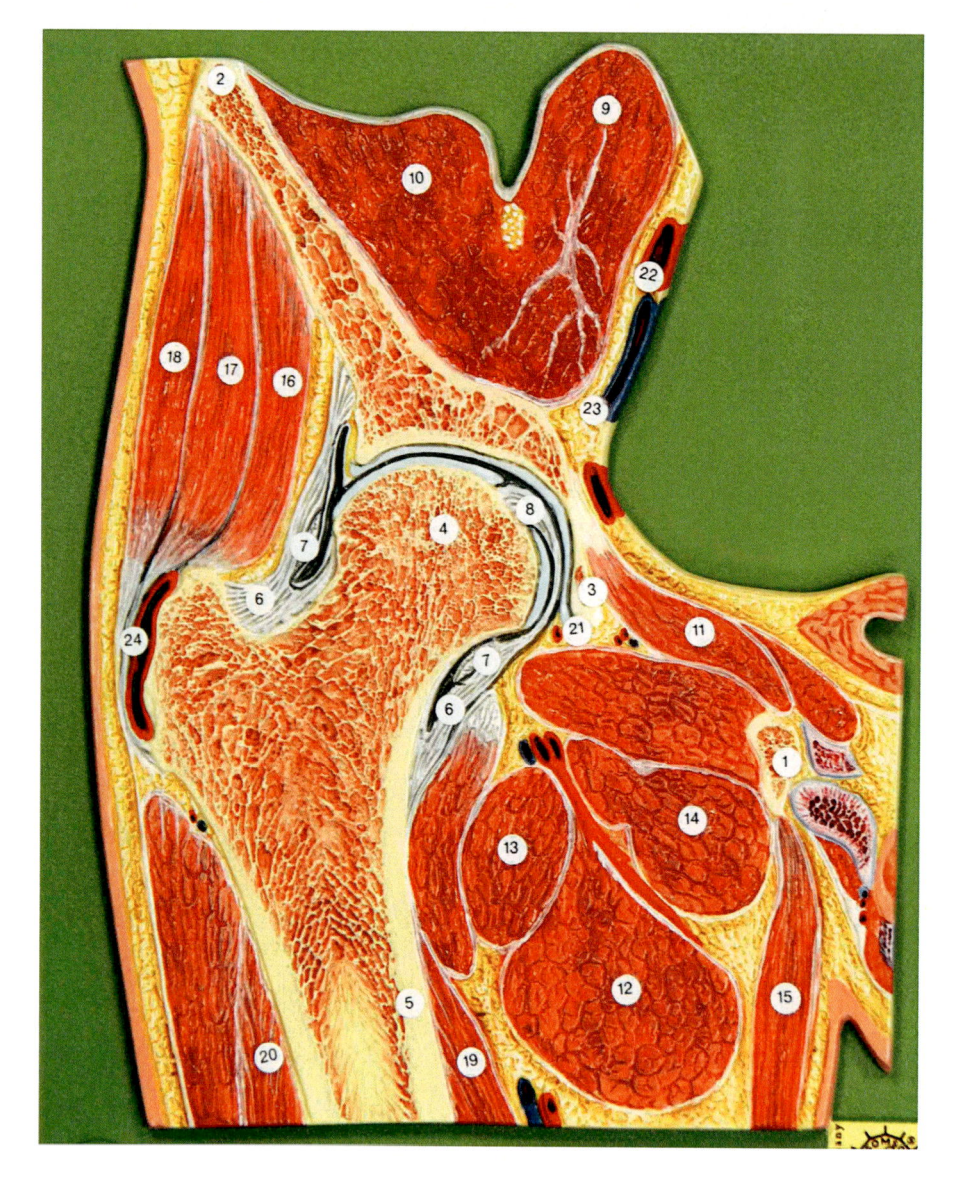

Figure 3-50. Hip Section
(Copyright Somso)

1. Pubis
2. Ilium
3. Ischium
4. Head of femur
5. Femur
6. Articular capsule
7. Annular fibers
8. Ligament of the head of the femur
9. Psoas major
10. Iliacus
11. Internal obturator
12. Adductor longus
13. Pectineus
14. Adductor brevis
15. Gracilis
16. Gluteus minimus
17. Gluteus medius
18. Tensor fasciae latae
19. Vastus medialis
20. Vastus lateralis
21. Adipose panniculus
22. Iliac vein
23. Iliac artery
24. Synovial bursa

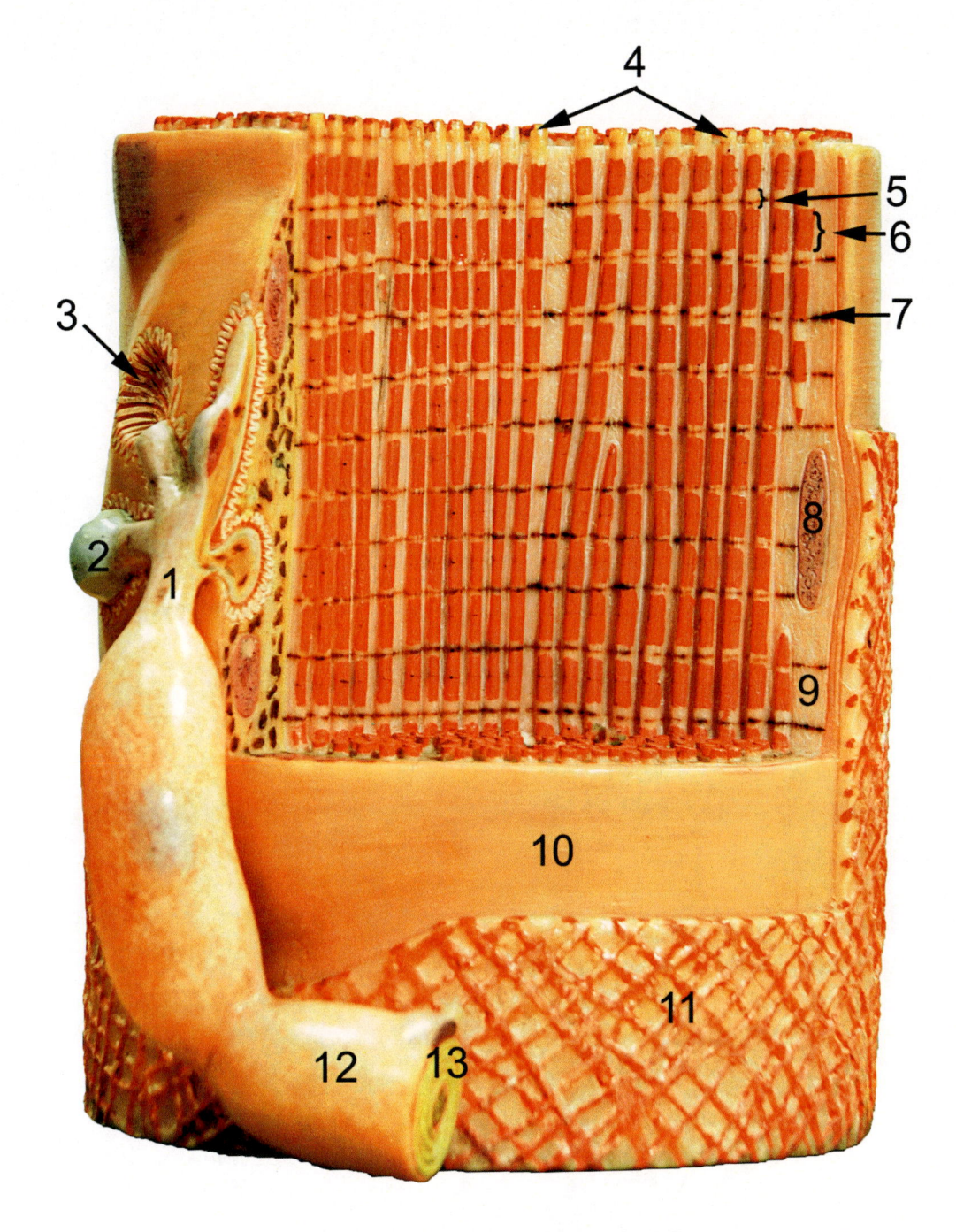

Figure 3-51. Skeletal Muscle Fiber
(Copyright Somso)

1. Somatic axon
2. Synaptic knob
3. Motor end plate
4. Myofibril
5. I band
6. A band
7. Z line
8. Nucleus of muscle fiber
9. Sarcoplasm
10. Sarcolemma
11. Endomysium
12. Neurolemma of Schwann cell
13. Myelin

Chapter 4
Nervous System Models

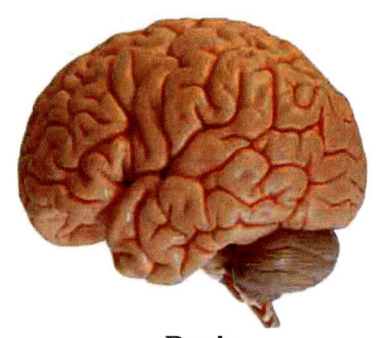

Brain
(Copyright Somso)

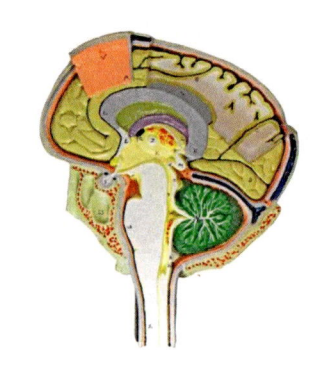

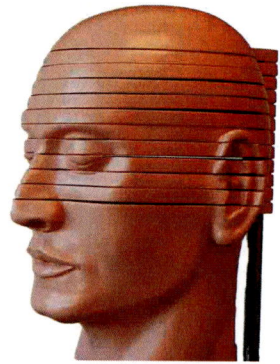

Head Axial sections
(Copyright Somso)

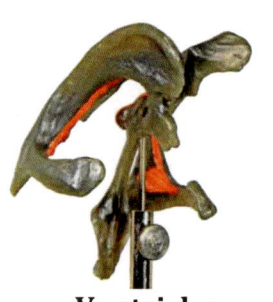

Sagittal Brain Section
(Copyright Denoyer-Geppert)

Ventricles
(Copyright Somso)

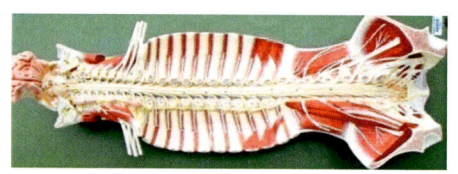

Spinal Cord in Spinal Canal
(Copyright Somso)

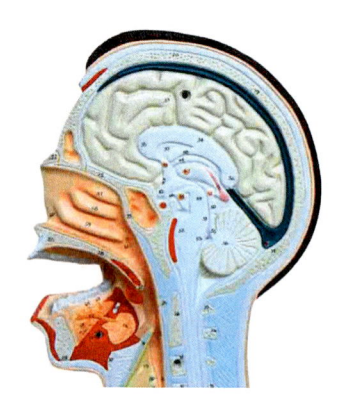

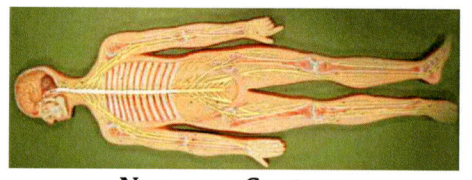

Head and Neck
(Copyright Denoyer-Geppert)

Nervous System
(Copyright Somso)

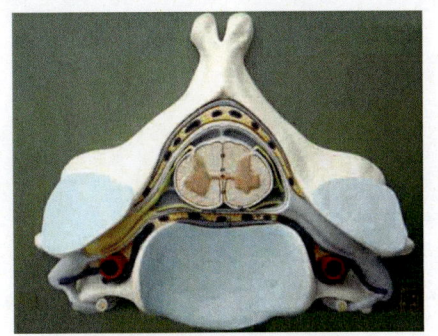

Fifth Cervical Vertebra - Spinal Cord
(Copyright Somso)

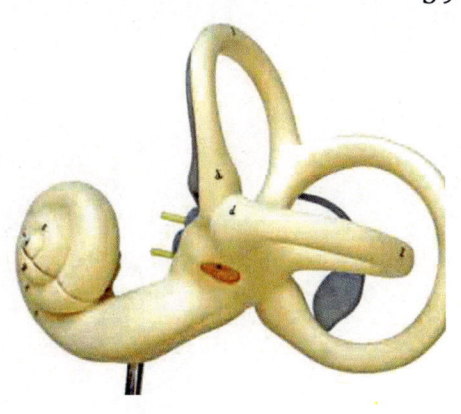

Inner Ear Labyrinth
(Copyright Somso)

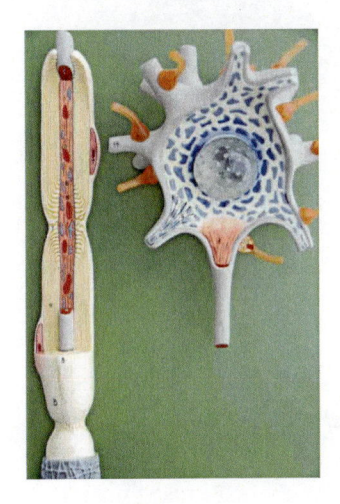

Neuron
(Copyright Somso)

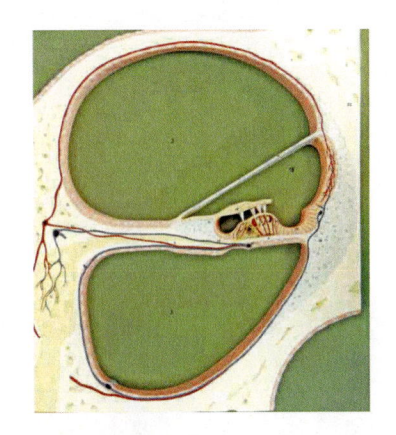

Cochlea section
(Copyright Somso)

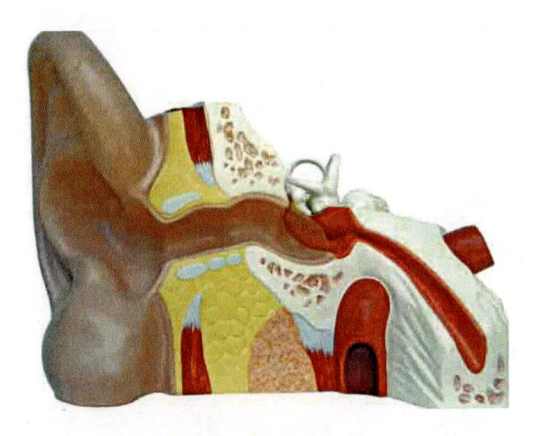

Ear
(Copyright 3B Scientific)

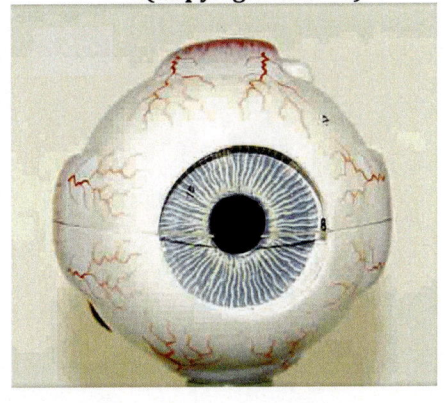

Eye
(Copyright Somso)

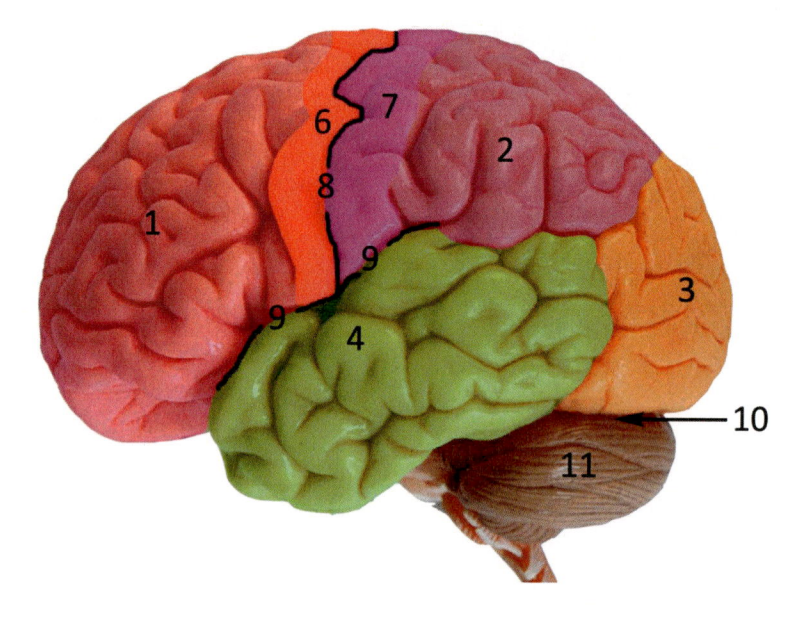

Figure 4-1. Brain - lateral

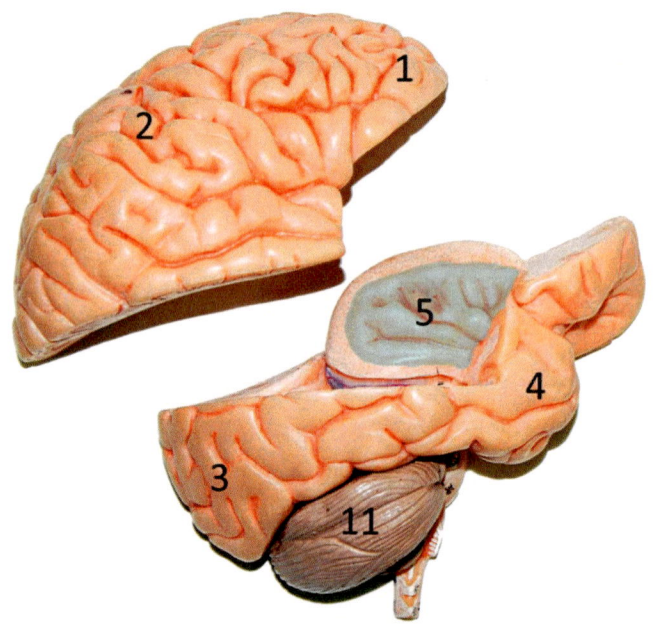

Figure 4-2. Brain - lateral (cut)
(Copyright Somso)

1. Frontal lobe
2. Parietal lobe
3. Occipital lobe
4. Temporal lobe
5. Insula
6. Precentral gyrus
7. Postcentral gyrus
8. Central sulcus
9. Lateral sulcus
10. Transverse fissure
11. Cerebellum

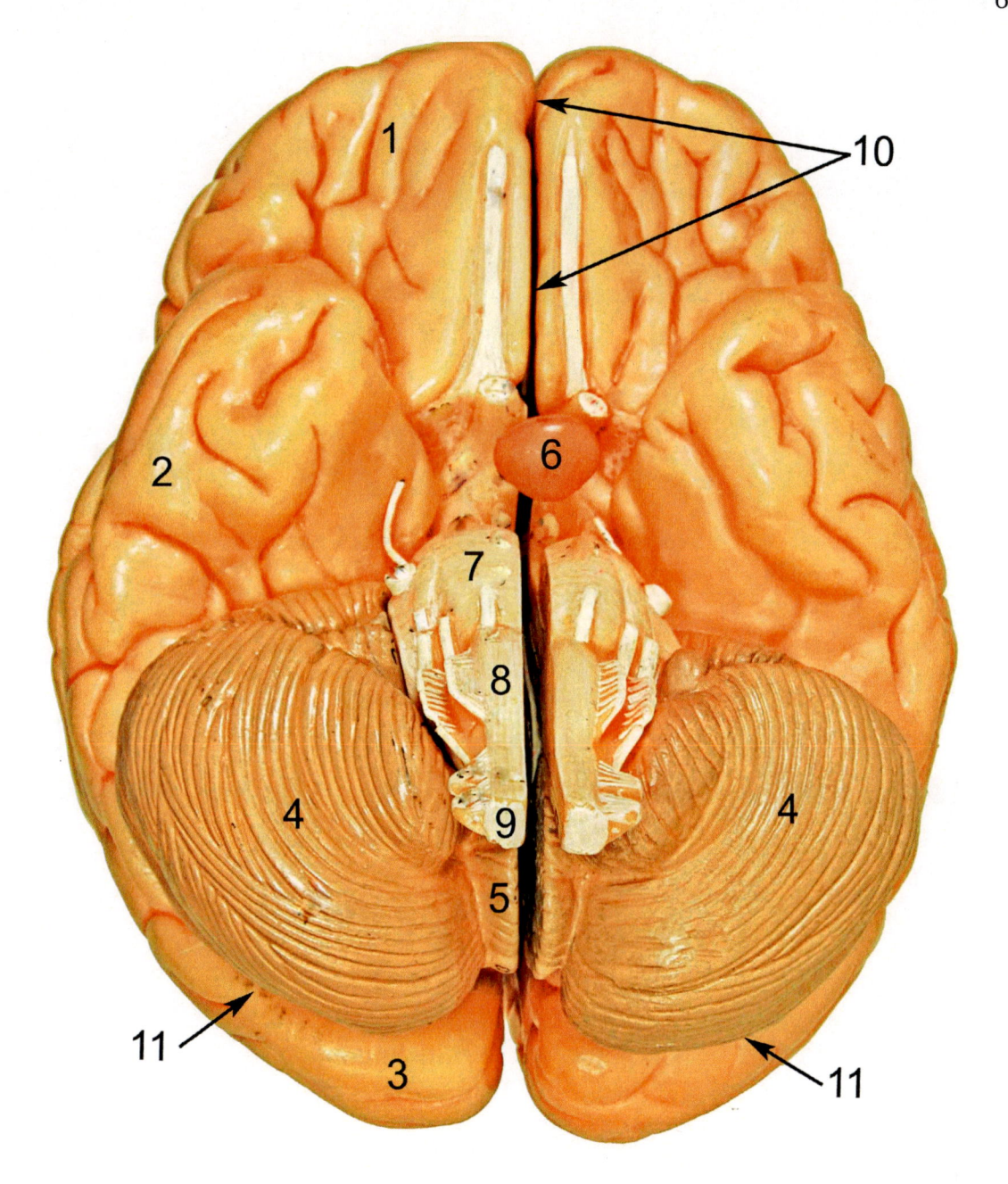

Figure 4-3. Brain - inferior
(Copyright Somso)

1. Frontal lobe of cerebrum
2. Temporal lobe of cerebrum
3. Occipital lobe of cerebrum
4. Cerebellar hemisphere
5. Vermis of cerebellum
6. Pituitary gland
7. Pons
8. Medulla oblongata
9. Spinal cord
10. Longitudinal fissure
11. Transverse fissure

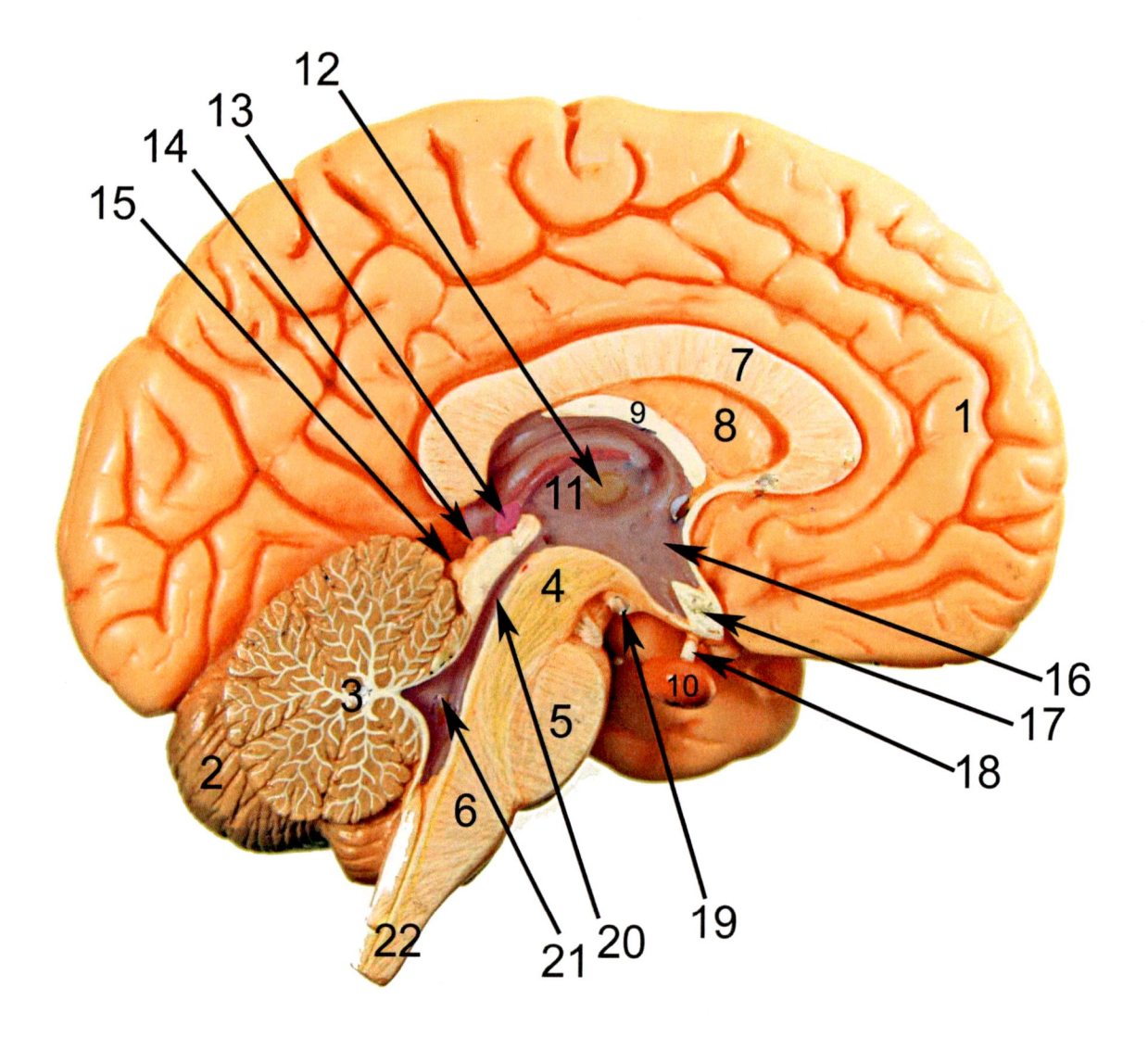

Figure 4-4. Brain - medial
(Copyright Somso)

1. Cerebrum
2. Cerebellum
3. Arbor vitae of cerebellum (white)
4. Midbrain
5. Pons
6. Medulla oblongata
7. Corpus callosum
8. Septum pellucidum
9. Fornix
10. Pituitary gland
11. Thalamus
12. Intermediate mass
13. Pineal gland
14. Superior colliculus
15. Inferior colliculus
16. Third ventricle (purple)
17. Optic chiasm (chiasma)
18. Infundibulum
19. Mamillary body
20. Aqueduct of midbrain (purple)
21. Fourth ventricle (purple)
22. Spinal cord

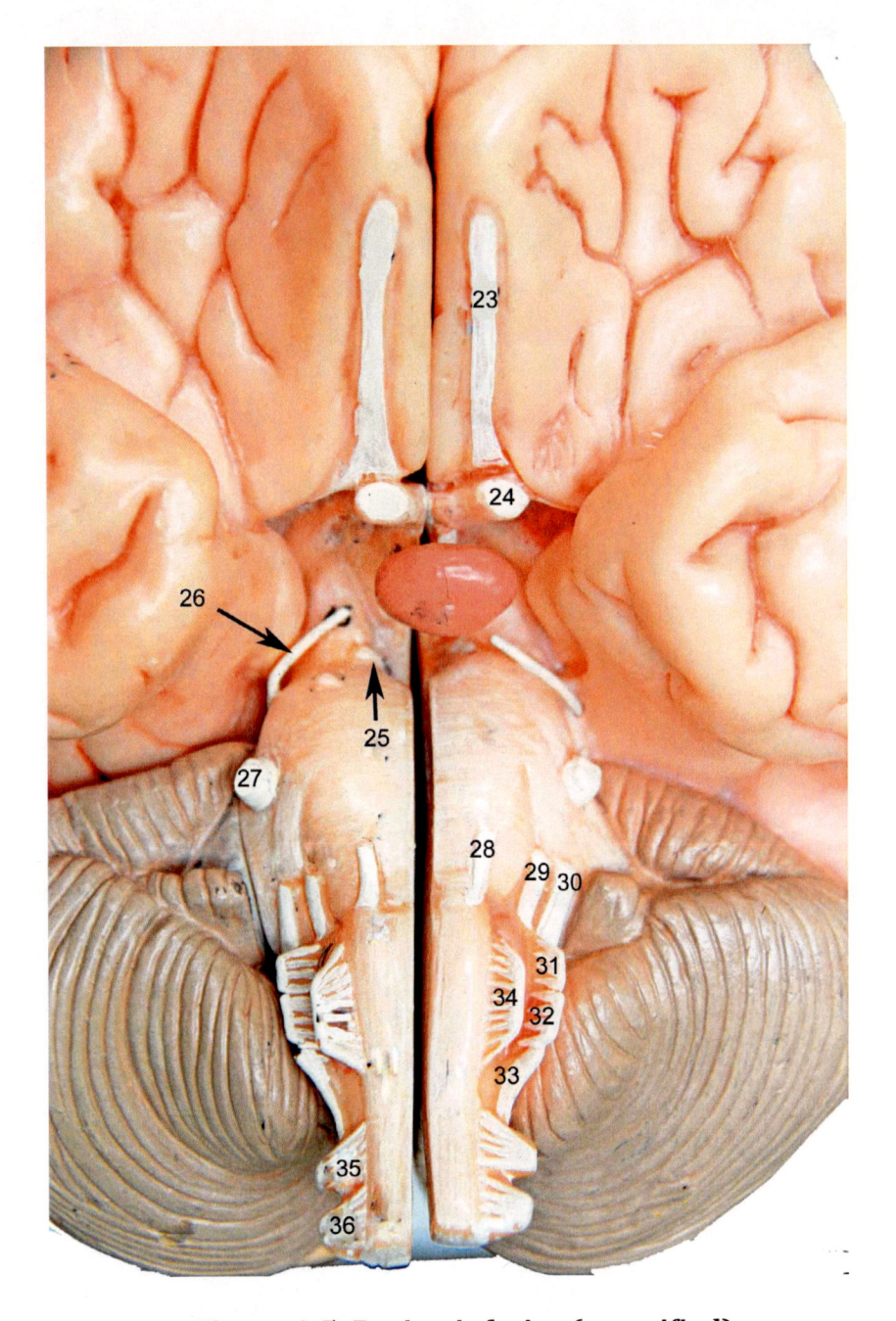

Figure 4-5. Brain - inferior (magnified)
(Copyright Somso)

23. Olfactory nerve (N I)
24. Optic nerve (N II)
25. Oculomotor nerve (N III)
26. Trochlear nerve (N IV)
27. Trigeminal nerve (N V)
28. Abducens nerve (N VI)
29. Facial nerve (N VII)
30. Vestibulocochlear nerve (N VIII)
31. Glossopharyngeal nerve (N IX)
32. Vagus nerve (N X)
33. Accessory nerve (N XI)
34. Hypoglossal nerve (N XII)
35. Cervical nerve 1 (C_1)
36. Cervical nerve 2 (C_2)

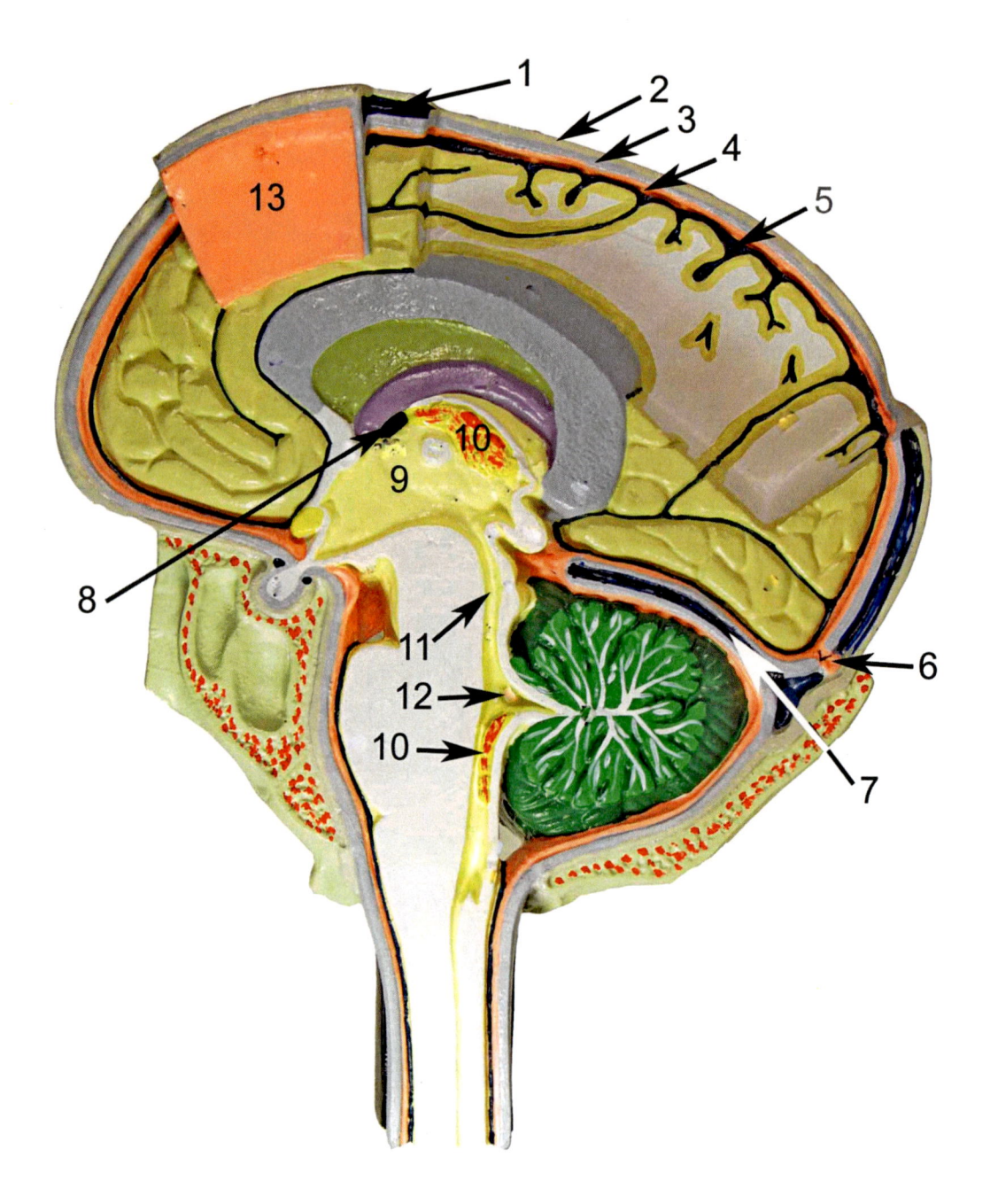

Figure 4-6. Brain section – medial
(Copyright Denoyer-Geppert)

1. Superior sagittal sinus
2. Endosteal layer of dura mater
3. Meningeal layer of dura mater
4. Arachnoid
5. Subarachnoid space
6. Arachnoid granulation
7. Straight sinus
8. Interventricular foramen
9. Third ventricle
10. Choroid plexus
11. Aqueduct of midbrain
12. Fourth ventricle
13. Arachnoid of wedge–like Falx cerebri

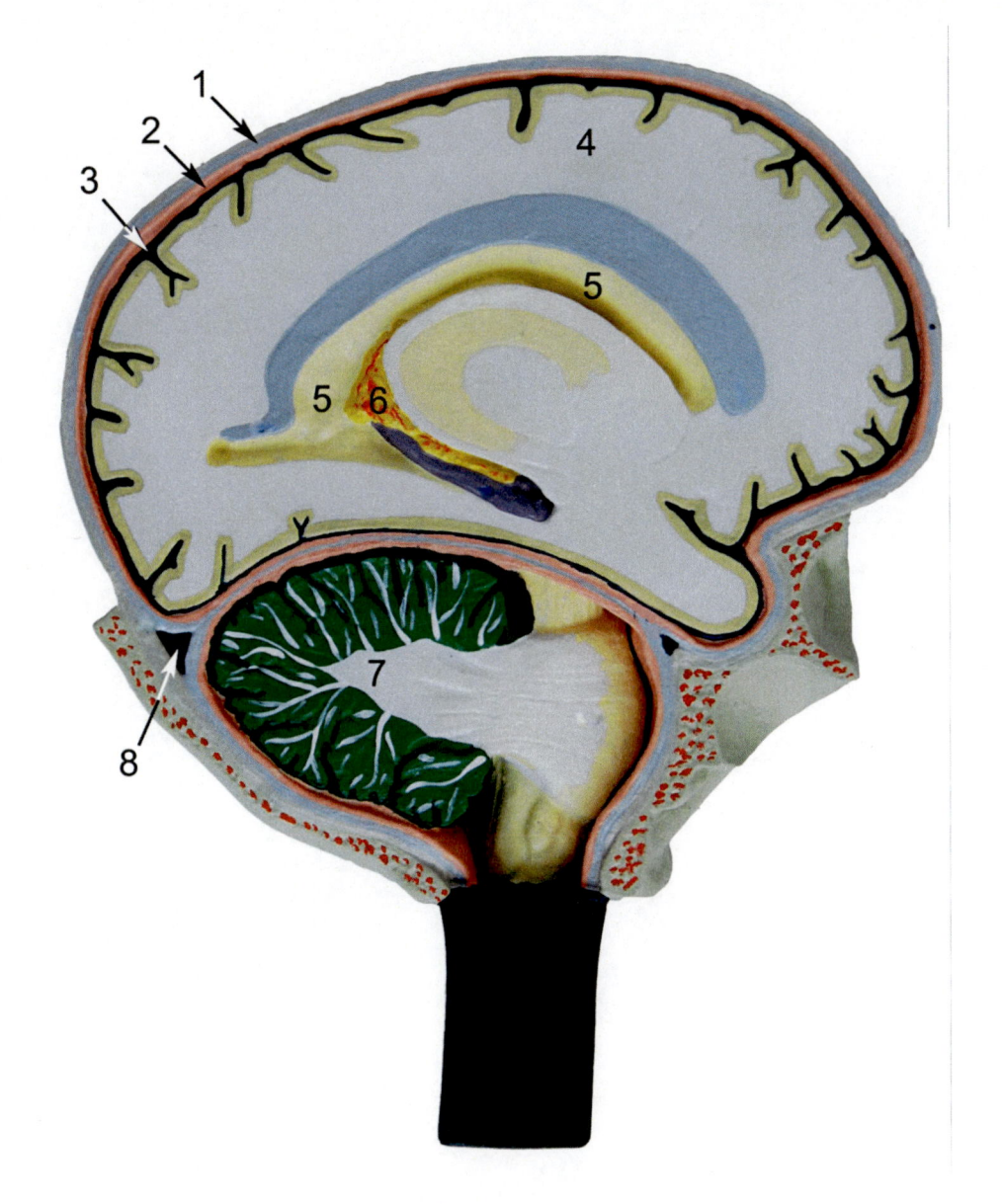

Figure 4-7. Brain section – lateral
(Copyright Denoyer-Geppert)

1. Dura mater
2. Arachnoid
3. Subarachnoid space
4. Cerebrum
5. Lateral ventricle
6. Choroid plexus
7. Cerebellum
8. Transverse sinus

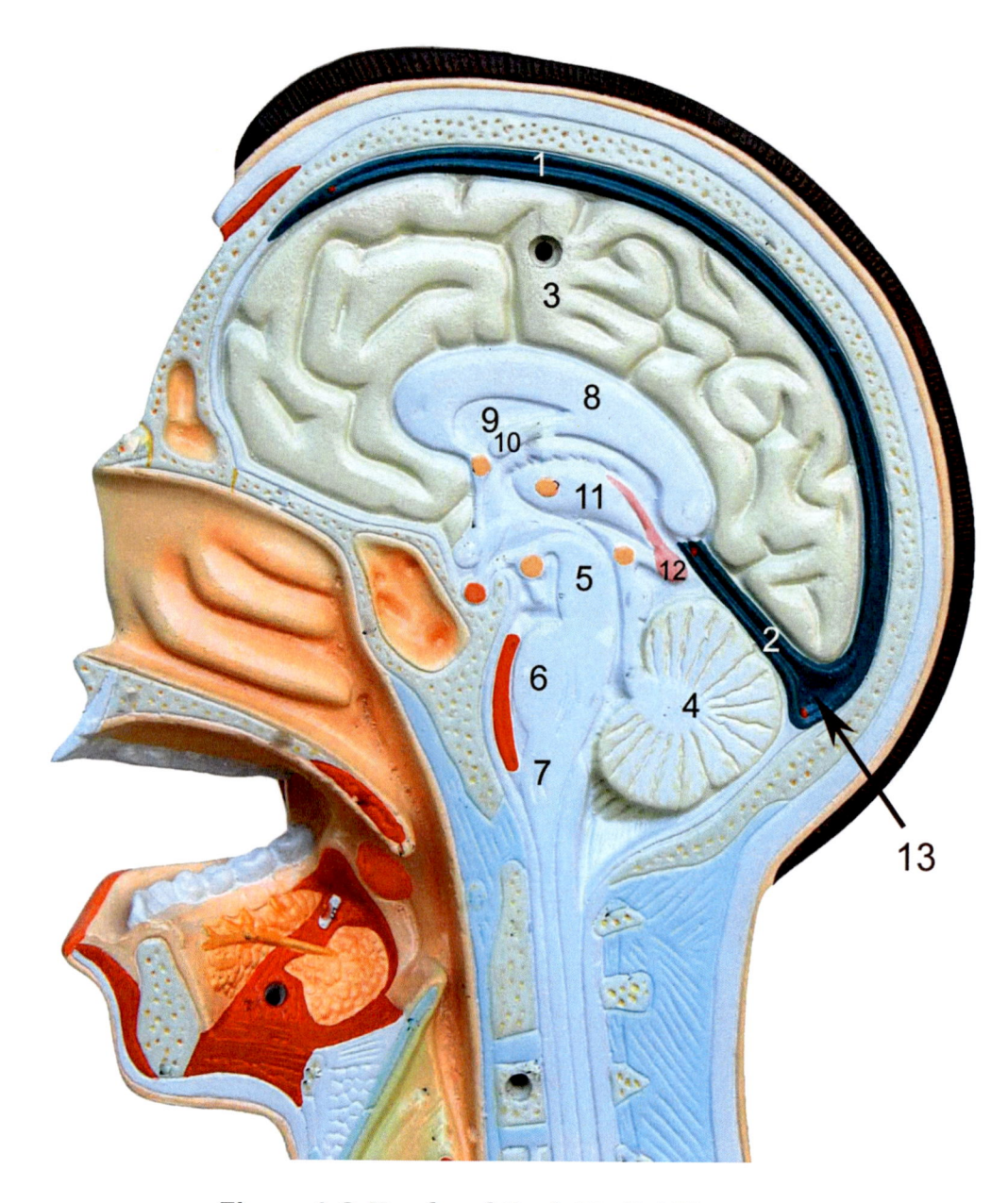

Figure 4-8. Head and Neck Medial View
(Copyright Denoyer-Geppert)

1. Superior sagittal sinus
2. Straight sinus
3. Cerebrum
4. Cerebellum
5. Midbrain
6. Pons
7. Medulla oblongata
8. Corpus callosum
9. Septum pellucidum
10. Fornix
11. Thalamus
12. Pineal gland
13. Confluence of sinuses

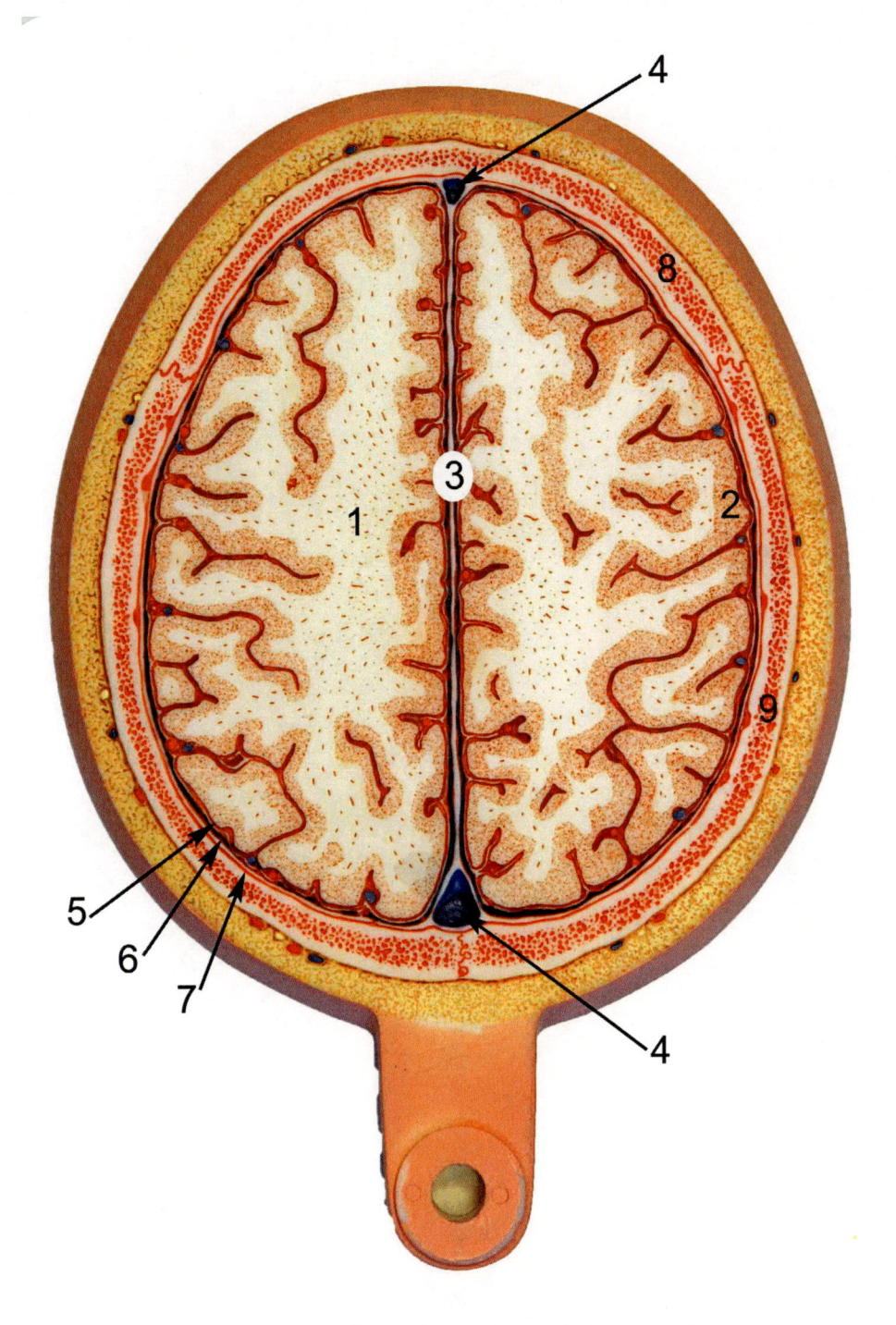

Figure 4-9. Brain – axial section level 1
(Copyright Somso)

1. Cerebral medulla (white matter)
2. Cerebral cortex (gray matter)
3. Falx cerebri in longitudinal fissure
4. Superior sagittal sinus
5. Pia mater
6. Subarachnoid space
7. Dura mater
8. Frontal bone
9. Parietal bone

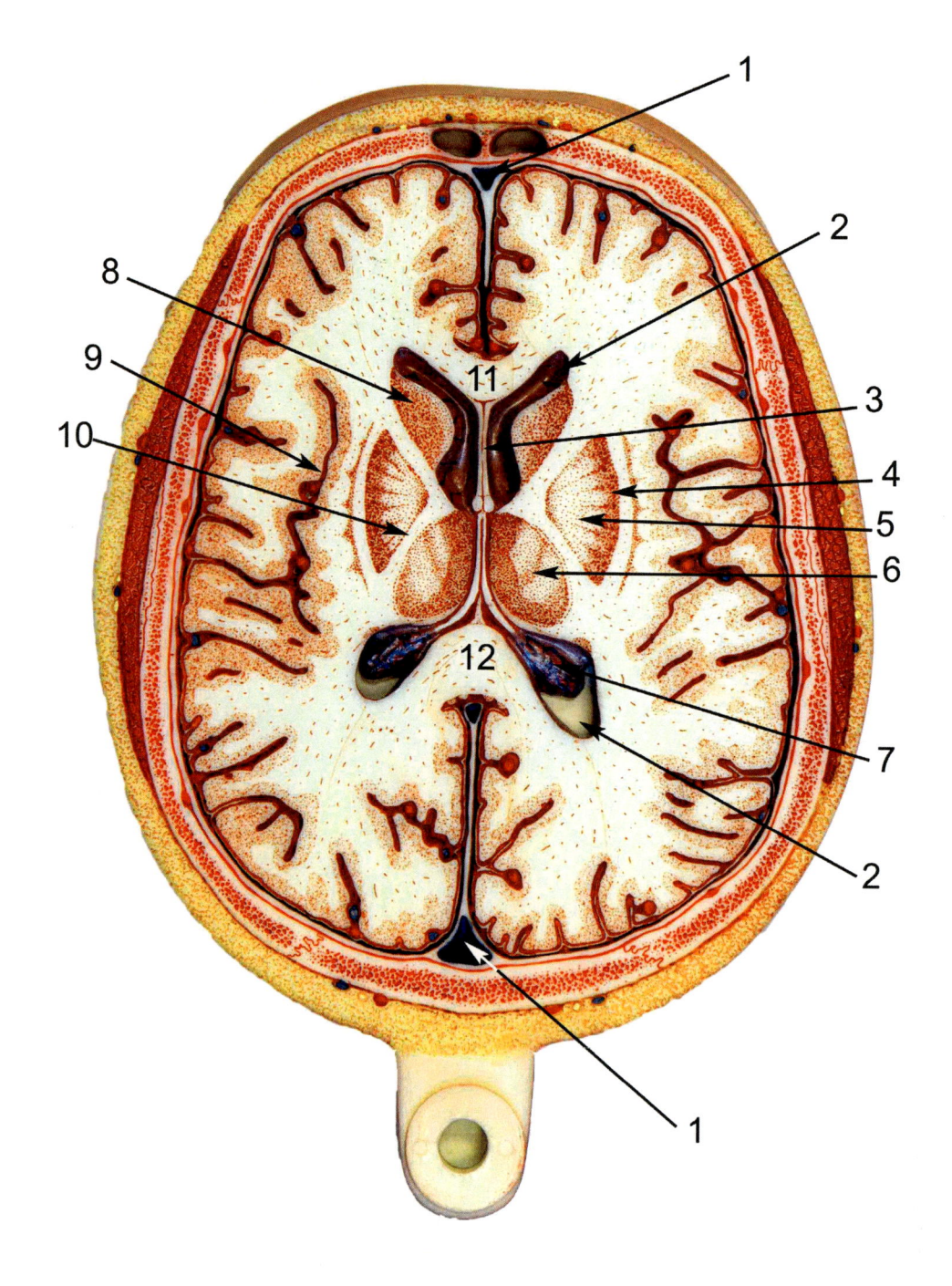

Figure 4-10. Brain – axial section level 4
(Copyright Somso)

1. Superior sagittal sinus
2. Lateral ventricle
3. Septum pellucidum
4. Globus pallidus
5. Putamen
6. Thalamus
7. Choroid plexus
8. Caudate nucleus
9. Insula
10. Internal capsule
11. Genu of corpus callosum
12. Splenium of corpus callosum

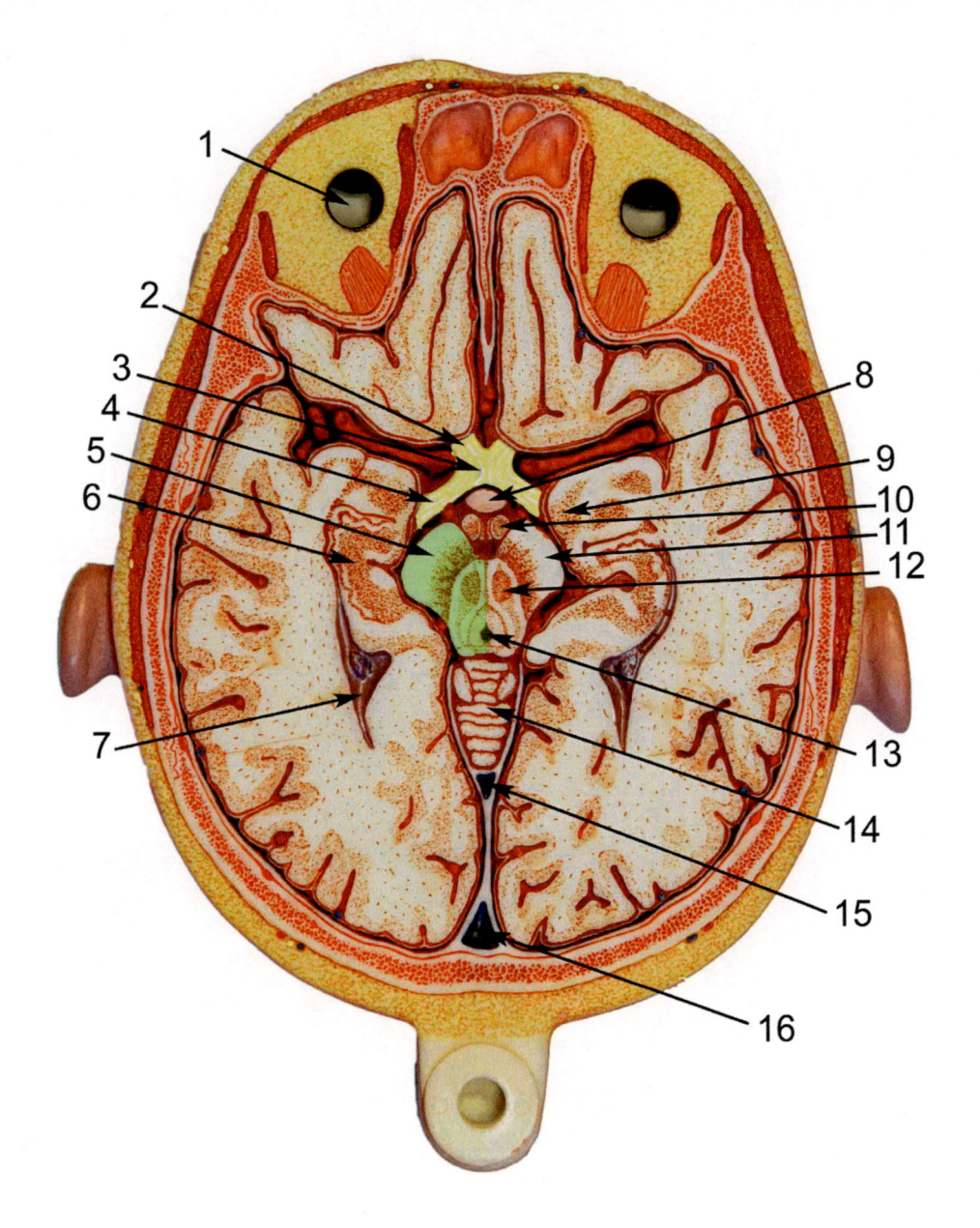

Figure 4-11. Brain – axial section level 6
(Copyright Somso)

1. Eyeball
2. Optic nerve (N II)
3. Optic chiasm (chiasma)
4. Optic tract
5. Midbrain (left side – green)
6. Hippocampus
7. Lateral ventricle
8. Infundibulum
9. Amygdala
10. Mamillary body
11. Cerebral peduncle of midbrain
12. Red nucleus of midbrain
13. Cerebral aqueduct
14. Vermis of cerebellum
15. Straight sinus
16. Superior sagittal sinus

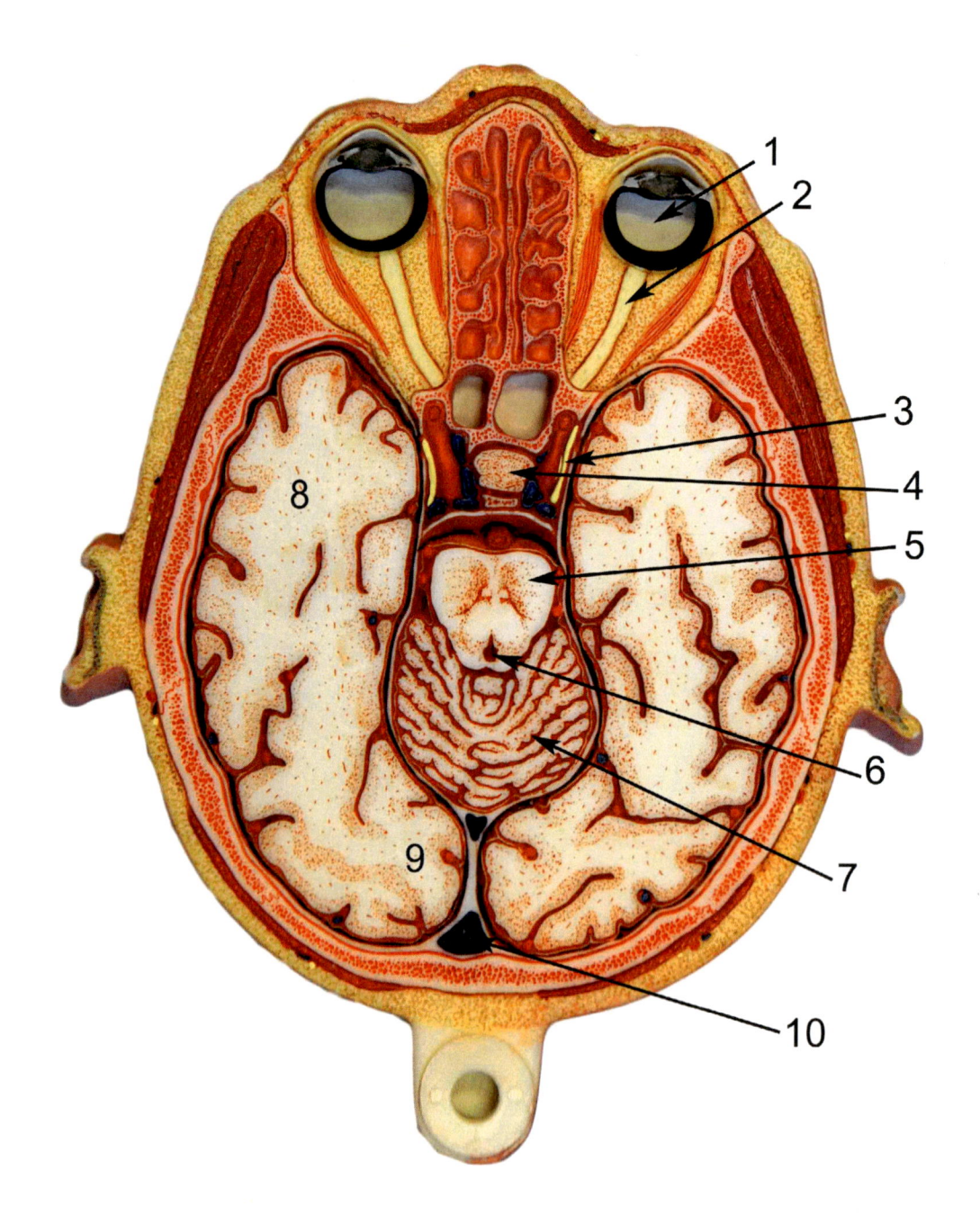

Figure 4-12. Brain – axial section level 7

1. Eyeball
2. Optic nerve (N II)
3. Oculomotor nerve (N III)
4. Pituitary gland (hypophysis)
5. Pons
6. Cerebral aqueduct
7. Cerebellum
8. Temporal lobe
9. Occipital lobe
10. Superior sagittal sinus

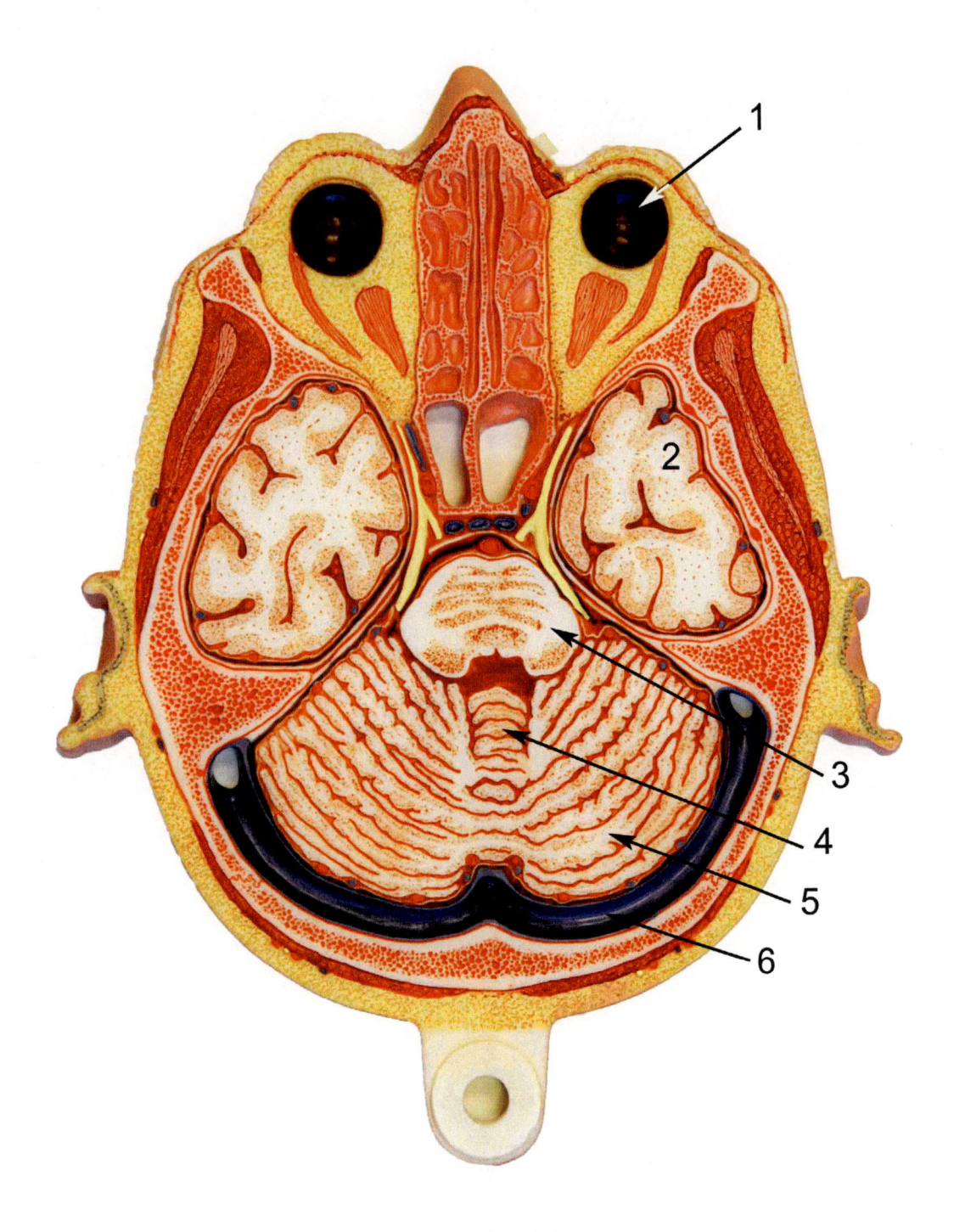

Figure 4-13. Brain – axial section level 8
(Copyright Somso)

1. Eyeball
2. Temporal lobe
3. Pons
4. Vermis of cerebellum
5. Cerebellar hemisphere
6. Transverse sinus

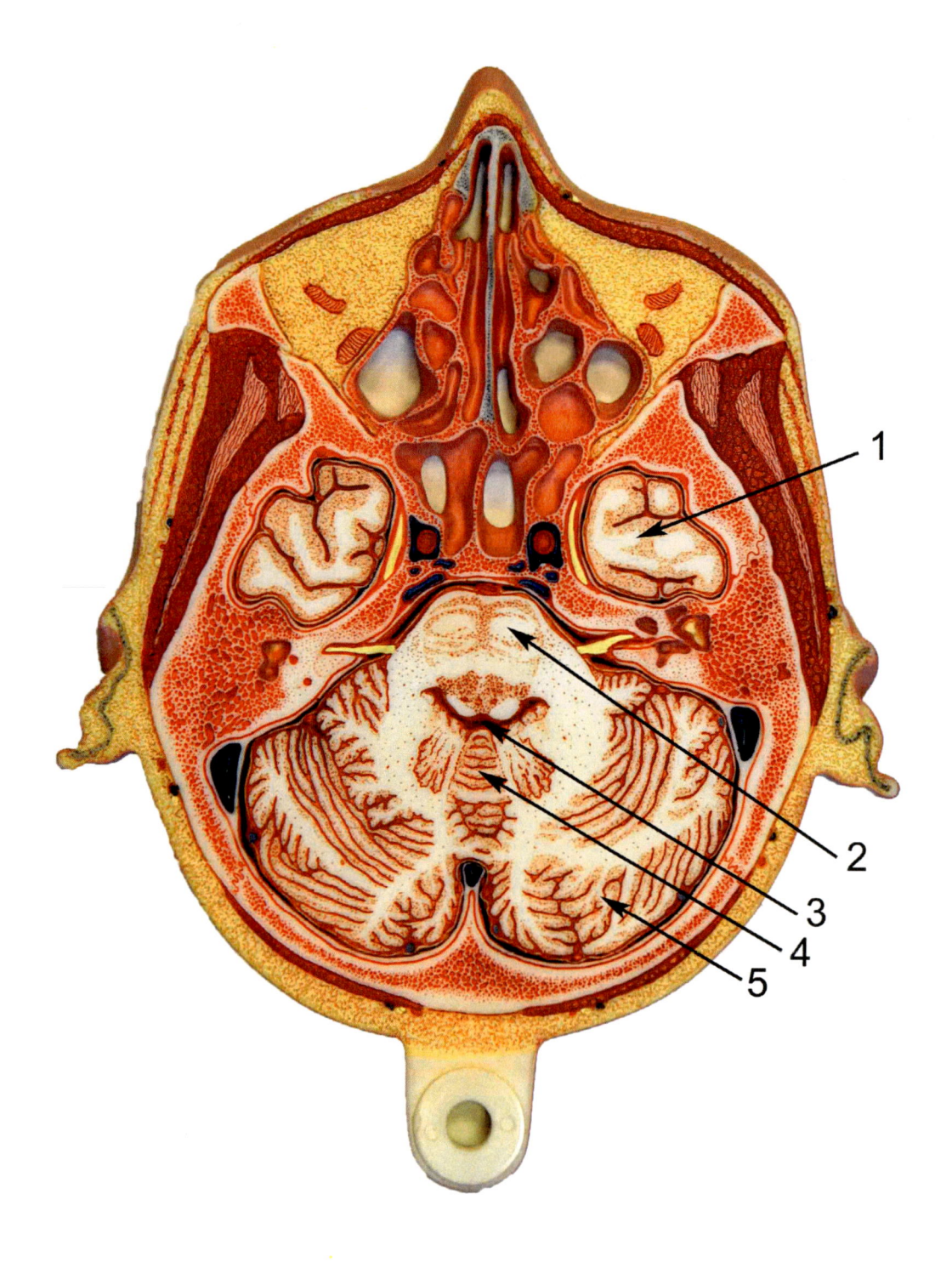

Figure 4-14. Brain – axial section level 9
(Copyright Somso)

1. Temporal lobe
2. Medulla oblongata
3. Fourth ventricle
4. Vermis of cerebellum
5. Cerebellar hemisphere

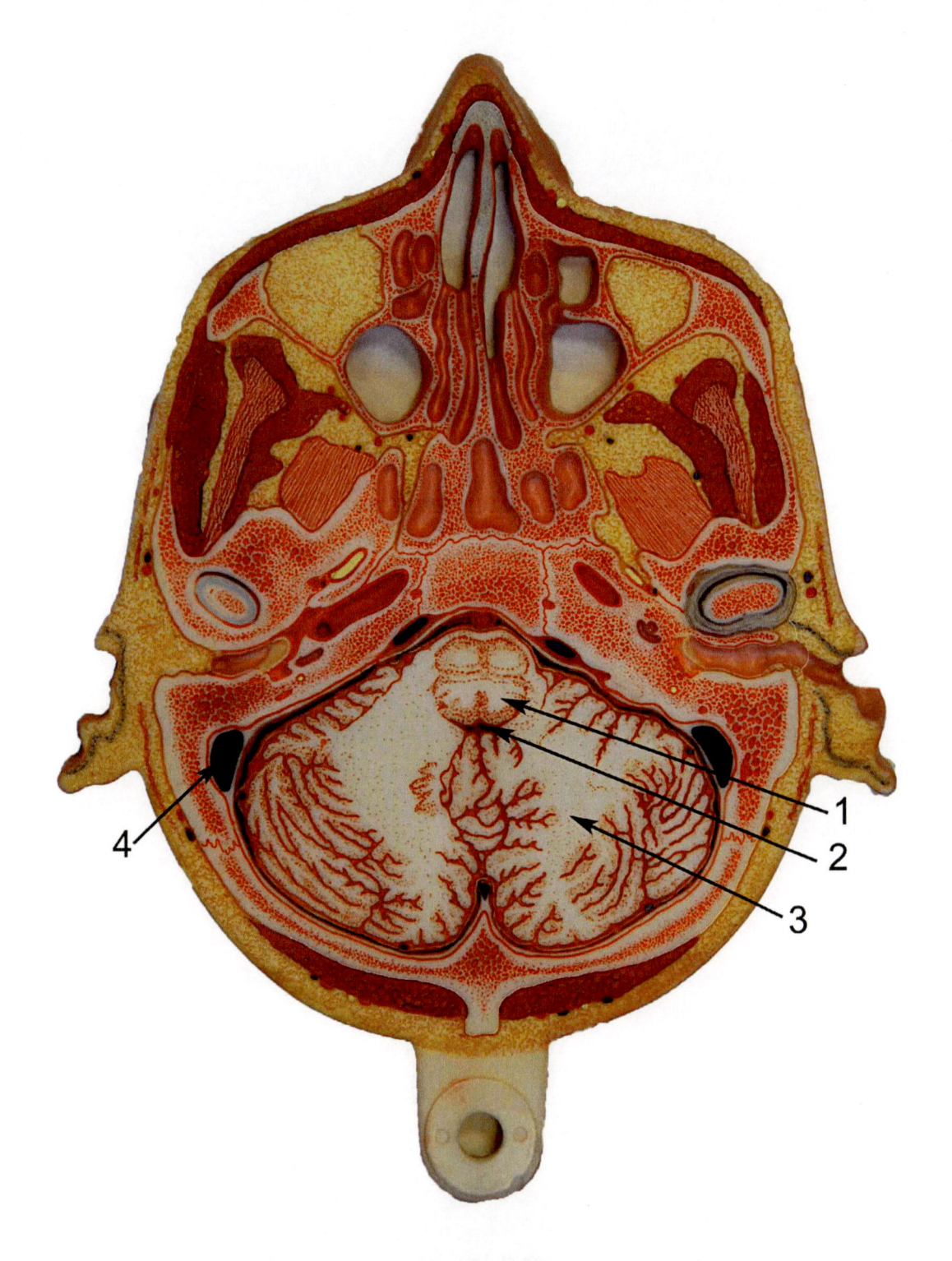

Figure 4-15. Brain – axial section level 10
(Copyright Somso)

1. Medulla oblongata
2. Fourth ventricle
3. Cerebellum
4. Sigmoid sinus

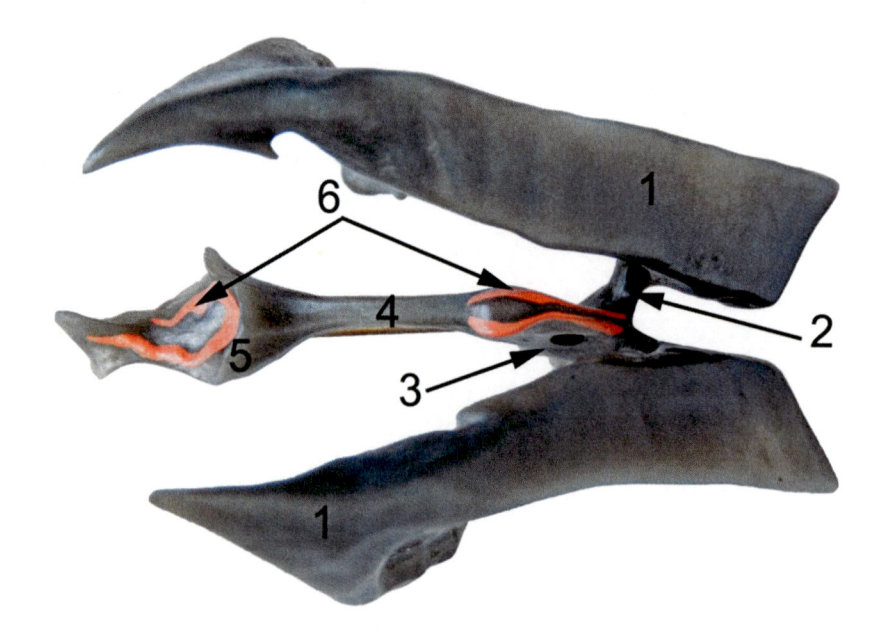

Figure 4-16. Ventricles – superior

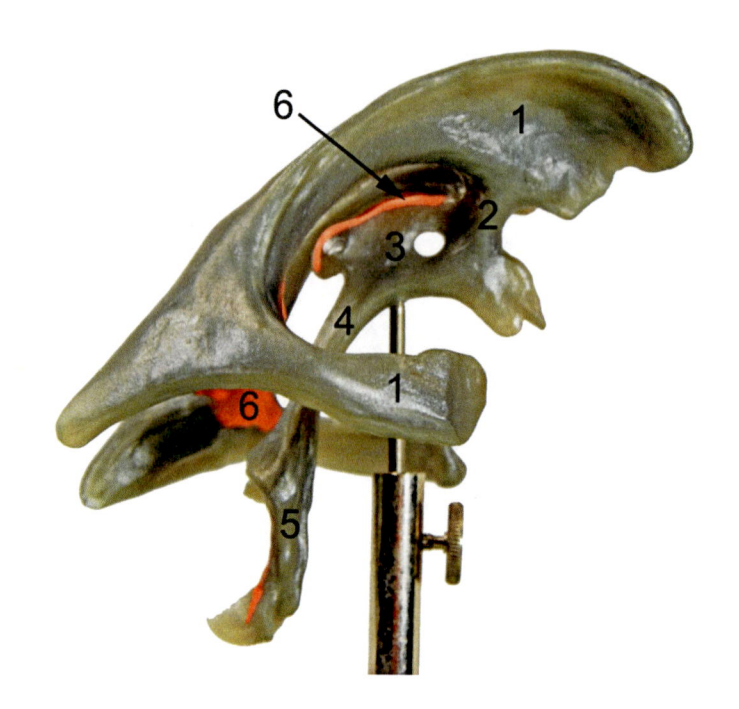

Figure 4-17. Ventricles - lateral
(Copyright Somso)

1. Lateral ventricles
2. Interventricular foramen
3. Third ventricle
4. Aqueduct of midbrain
5. Fourth ventricle
6. Choroid plexus

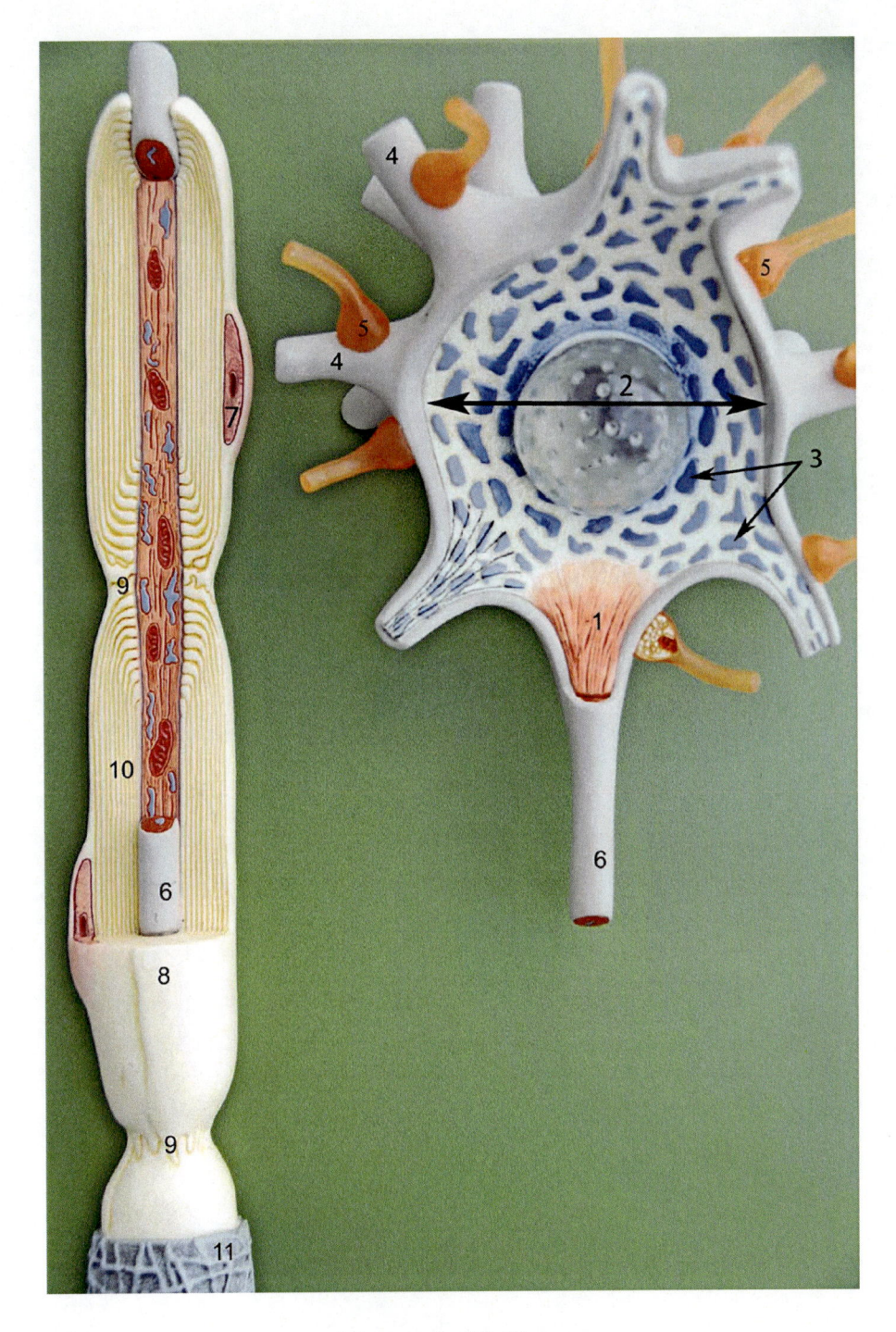

Figure 4-18. Neuron
(Copyright Somso)

1. Axon hillock
2. Cell body
3. Nissl bodies
4. Dendrite
5. Axon terminal
6. Axon
7. Nucleus of Schwann cell
8. Neurolemma
9. Node of Ranvier (myelin sheath gap)
10. Myelin sheath
11. Endoneurium

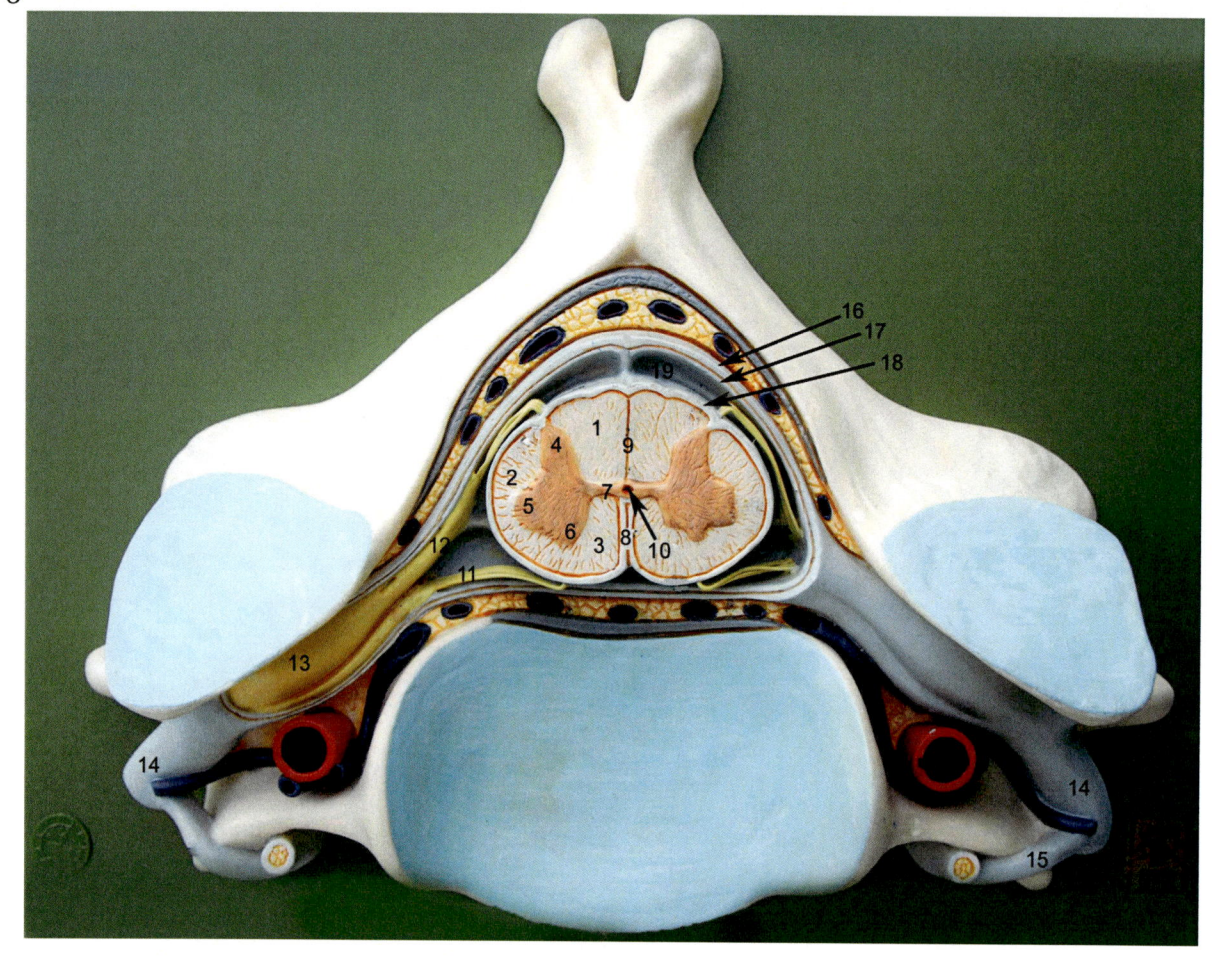

Figure 4-19. Spinal Cord and Nerve Branches – cross section
(Copyright Somso)

1. Posterior white column
2. Lateral white column
3. Anterior white column
4. Posterior gray horn
5. Lateral gray horn
6. Anterior gray horn
7. Gray commissure
8. Anterior median fissure
9. Posterior median sulcus
10. Central canal
11. Anterior (ventral) root
12. Posterior (dorsal) root
13. Posterior (dorsal) root ganglion
14. Ventral (anterior) ramus
15. Rami communicantes
16. Dura mater
17. Arachnoid
18. Pia mater
19. Subarachnoid space

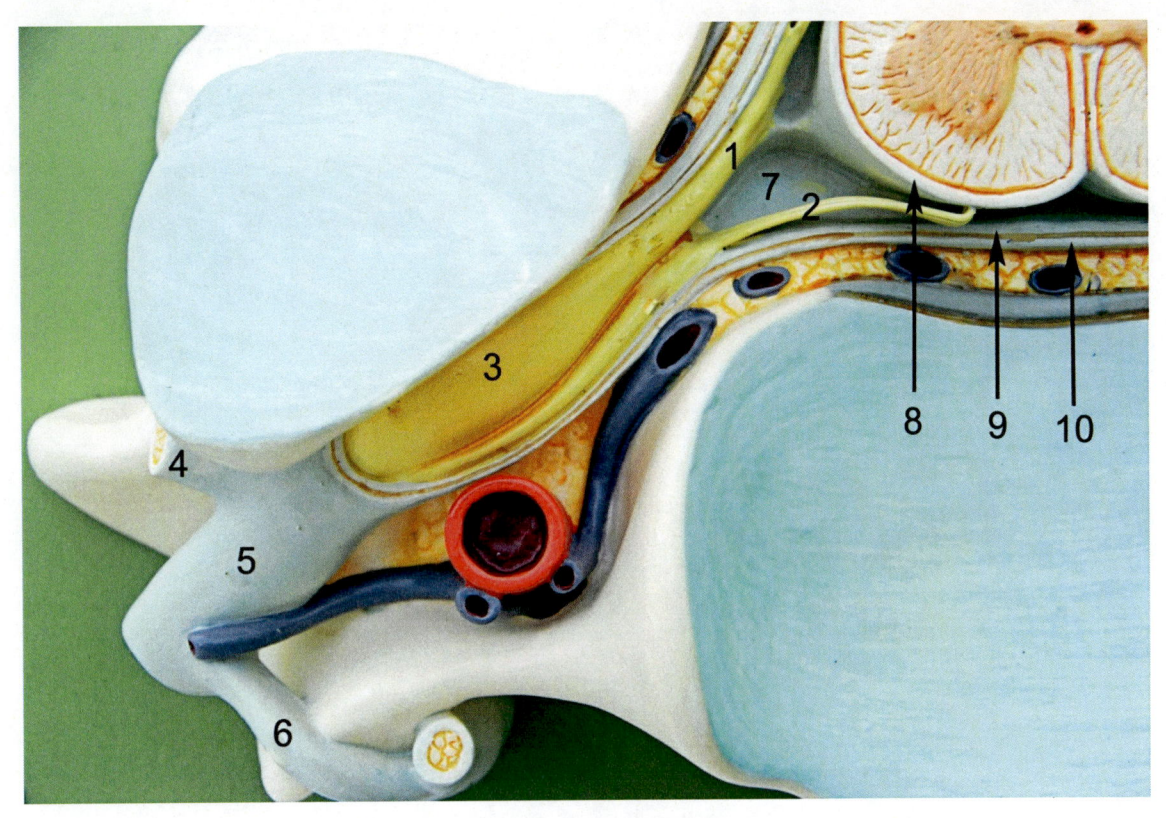

Figure 4-20. Spinal Cord and Nerve Branches
(Copyright Somso)

1. Posterior (dorsal) root
2. Anterior (ventral) root
3. Posterior (dorsal) root ganglion
4. Dorsal (posterior) ramus
5. Ventral (anterior) ramus
6. Rami communicantes
7. Subarachnoid space
8. Pia mater
9. Arachnoid
10. Dura mater

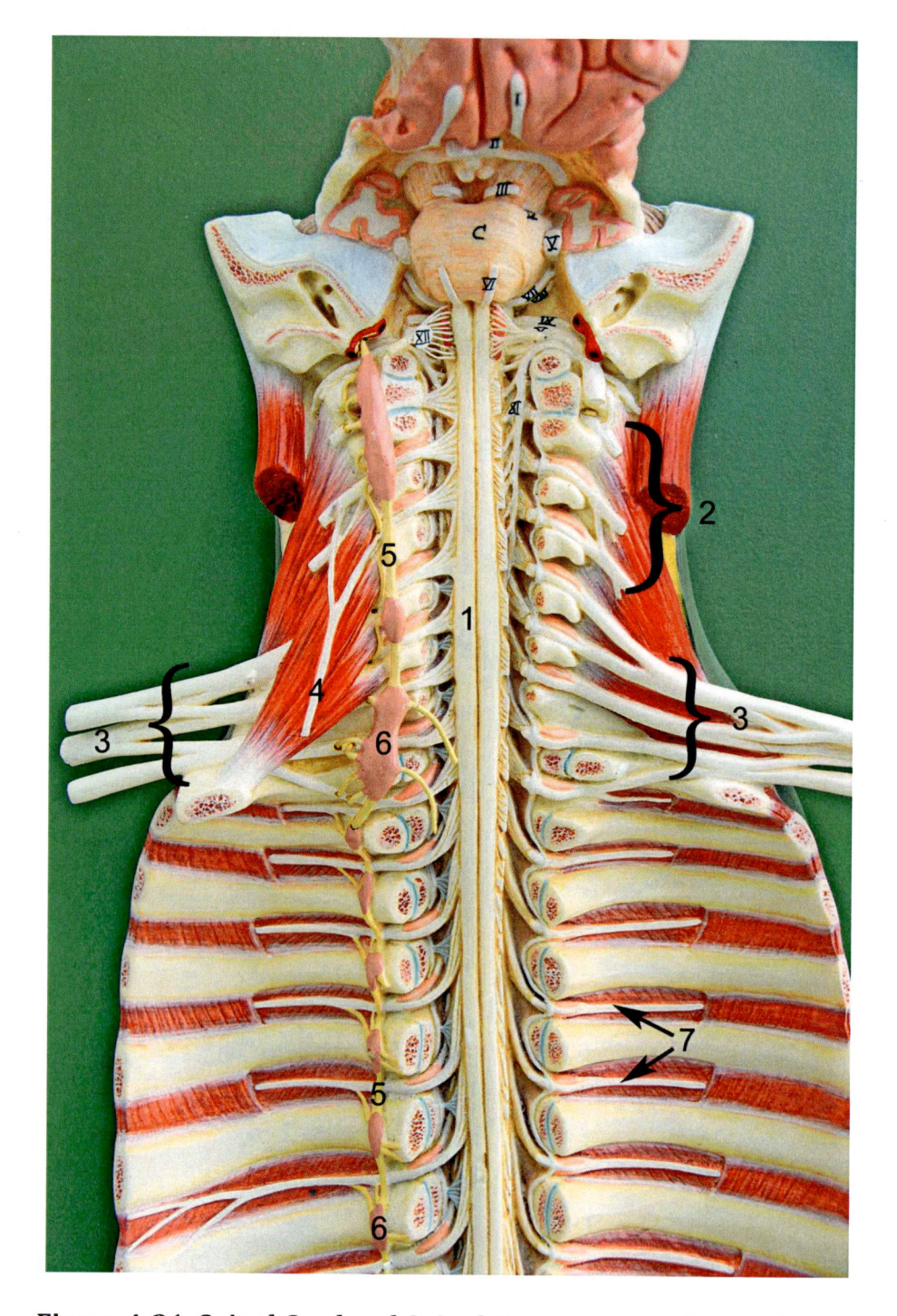

Figure 4-21. Spinal Cord and Spinal Nerves – superior section
(Copyright Somso)

1. Cervical enlargement
2. Cervical plexus
3. Brachial plexus
4. Phrenic nerve
5. Sympathetic chain (trunk)
6. Sympathetic ganglion
7. Intercostal nerve

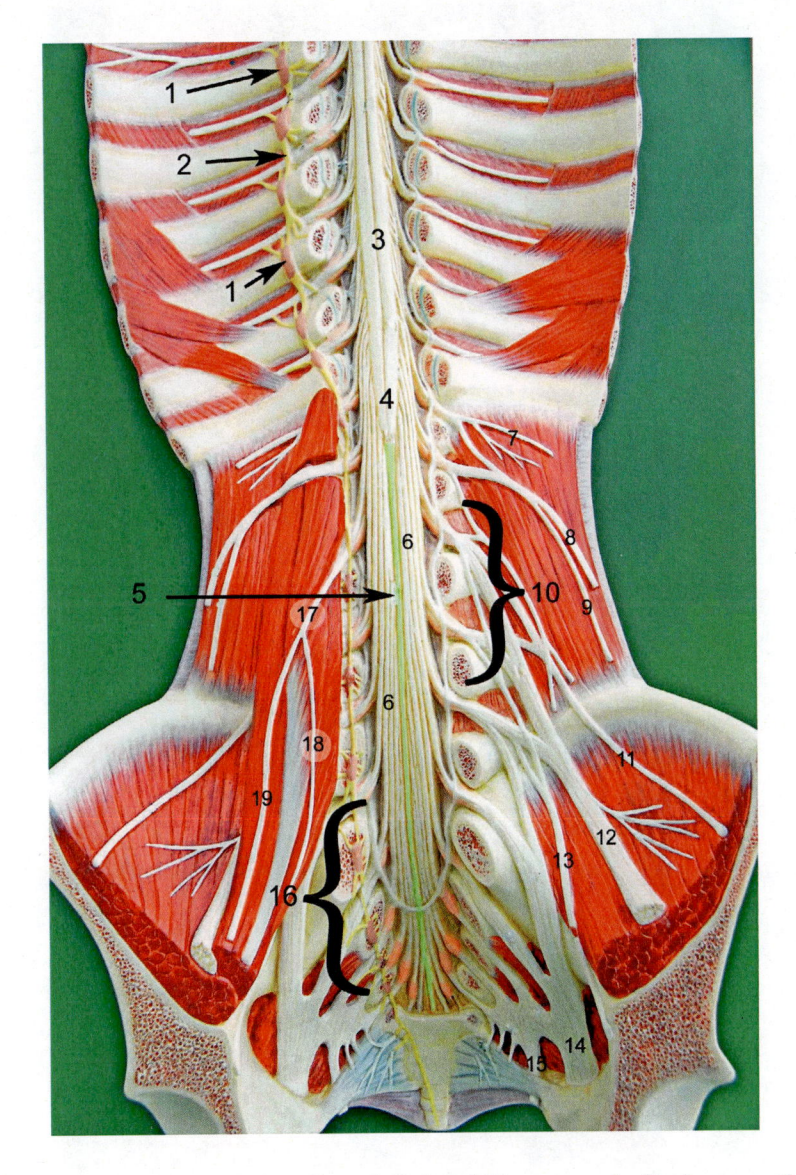

Figure 4-22. Spinal Cord and Spinal Nerves – inferior section
(Copyright Somso)

1. Sympathetic ganglion
2. Sympathetic chain (trunk)
3. Lumbar enlargement
4. Conus medullaris
5. Filum terminale
6. Cauda equina
7. Subcostal nerve
8. Iliohypogastric nerve
9. Ilioinguinal nerve
10. Lumbar plexus
11. Lateral femoral cutaneous nerve
12. Femoral nerve
13. Obturator nerve
14. Sciatic nerve
15. Pudendal nerve
16. Sacral plexus
17. Genitofemoral nerve
18. Femoral branch of genitofemoral nerve
19. Genital branch of genitofemoral nerve

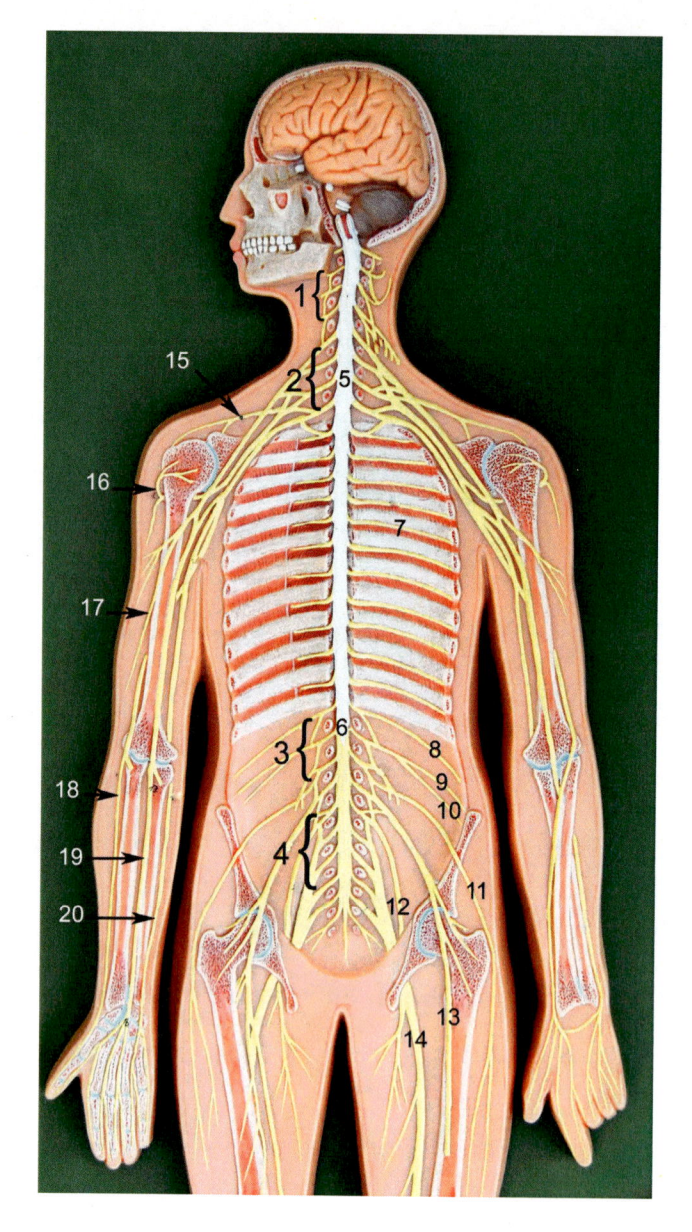

Figure 4-23. Nervous System – upper section
(Copyright Somso)

1. Cervical plexus
2. Brachial plexus
3. Lumbar plexus
4. Sacral plexus
5. Cervical enlargement of spinal cord
6. Conus medullaris
7. Intercostal nerve
8. Subcostal nerve
9. Iliohypogastric nerve
10. Ilioinguinal nerve
11. Lateral femoral cutaneous nerve
12. Obturator nerve
13. Femoral nerve
14. Sciatic nerve
15. Suprascapular nerve
16. Axillary nerve
17. Musculocutaneous nerve
18. Radial nerve
19. Median nerve
20. Ulnar nerve

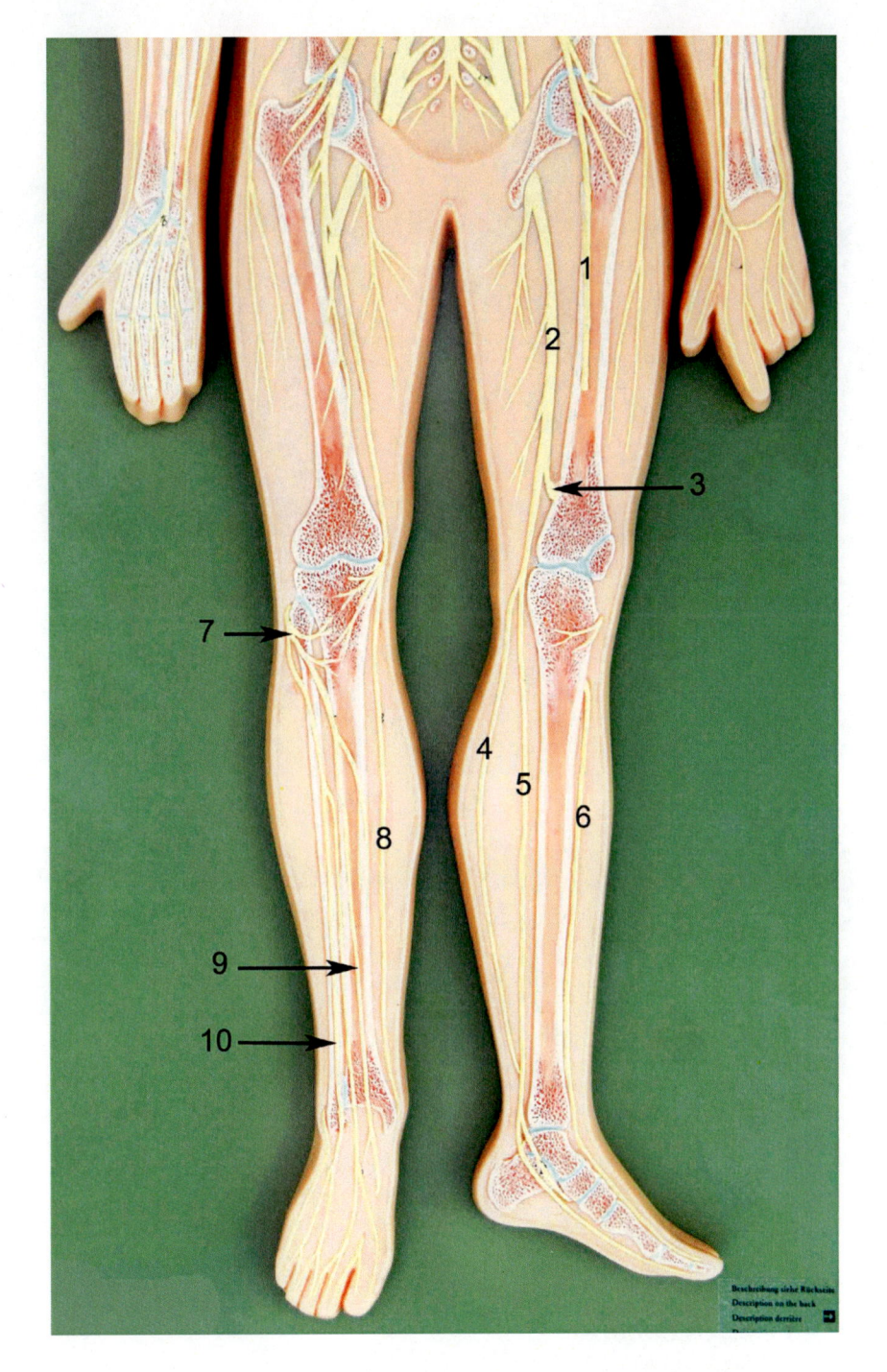

Figure 4-24. Nervous System – lower section
(Copyright Somso)

1. Femoral nerve
2. Sciatic nerve
3. Common fibular (peroneal) nerve
4. Sural nerve
5. Tibial nerve
6. Deep fibular (peroneal) nerve
7. Common fibular (peroneal) nerve
8. Saphenous nerve
9. Deep fibular (peroneal) nerve
10. Superficial fibular (peroneal) nerve

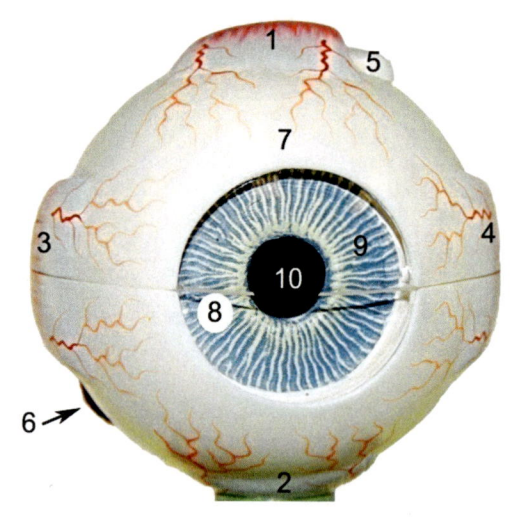

Figure 4-25. Eye - anterior

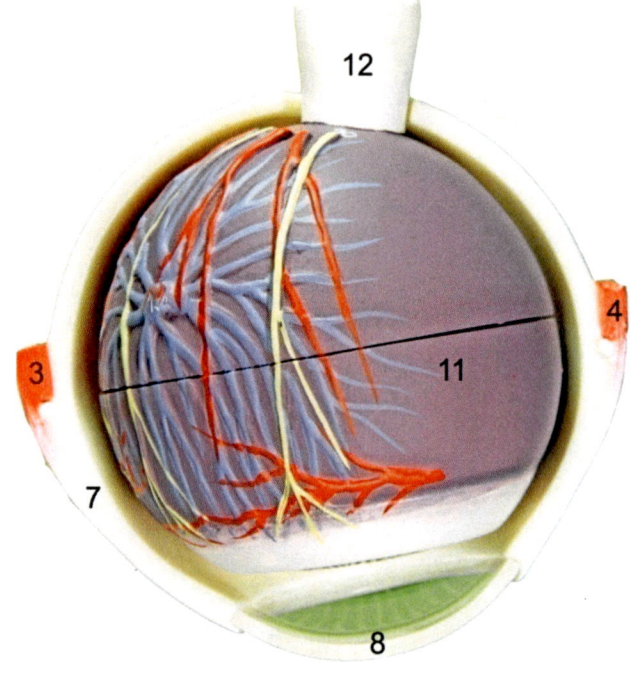

Figure 4-27. Eye Interior - superior

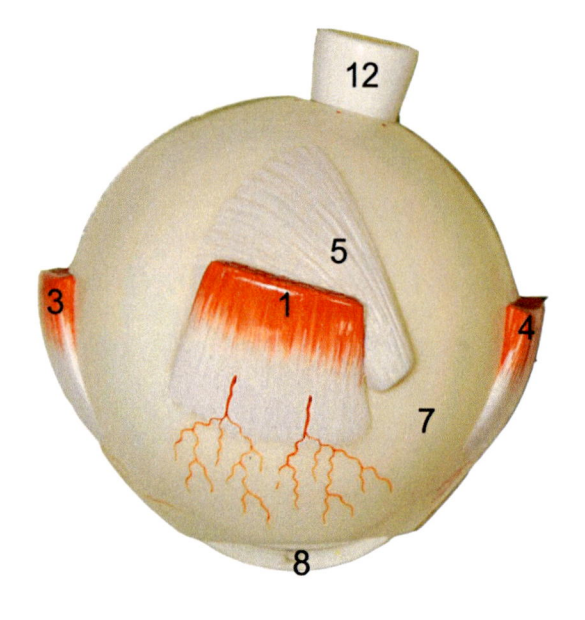

Figure 4-26. Eye - superior
(Copyright Somso)

1. Superior rectus muscle
2. Inferior rectus muscle
3. Lateral rectus muscle
4. Medial rectus muscle
5. Superior oblique (tendon)
6. Inferior oblique muscle
7. Sclera
8. Cornea
9. Iris
10. Pupil
11. Choroid
12. Optic nerve

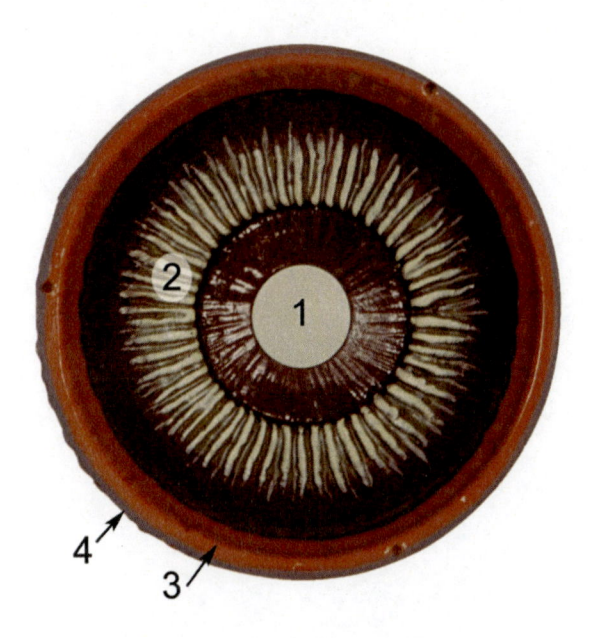

Figure 4-28. Eye Interior - anterior

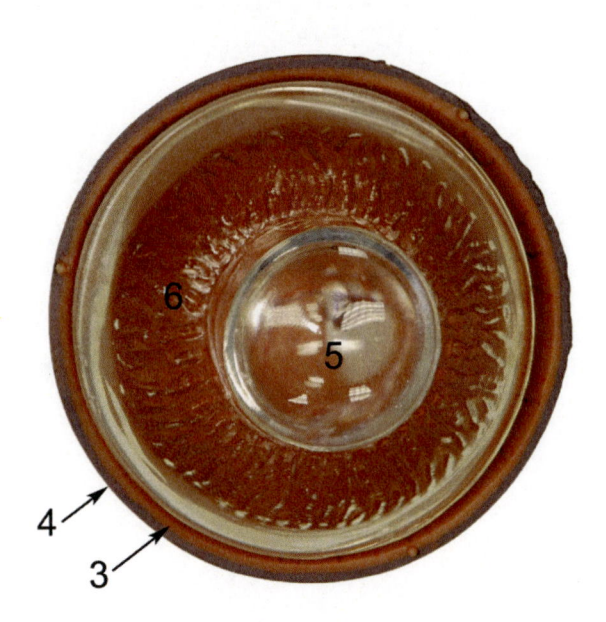

Figure 4-29. Eye Interior - vitreous chamber

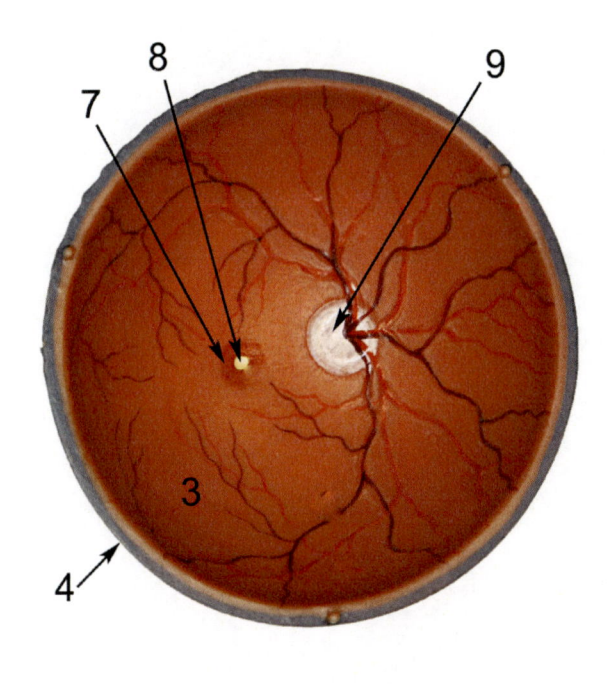

Figure 4-30. Eye Interior – posterior
(Copyright Somso)

1. Pupil
2. Suspensory ligaments
3. Retina
4. Choroid
5. Lens
6. Vitreous humor (body)
7. Macula lutea
8. Fovea centralis
9. Optic disk

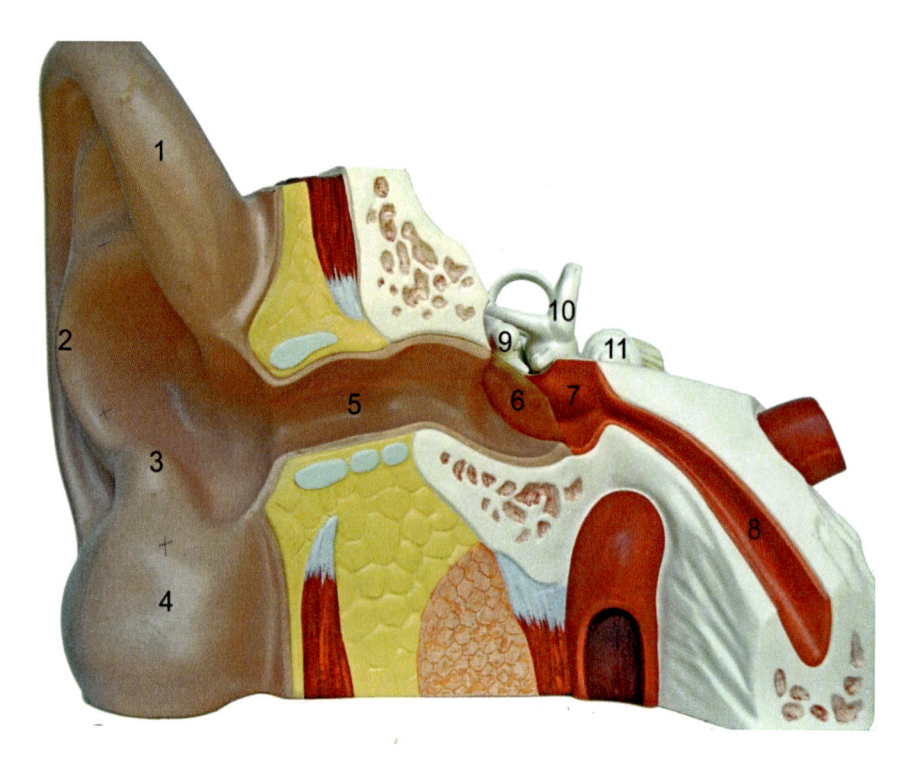

Figure 4-31. Ear
(Copyright 3B Scientific)

1. Helix of auricle (pinna)
2. Antihelix of auricle
3. Antitragus of auricle
4. Lobe of auricle
5. External acoustic meatus
6. Tympanic membrane
7. Tympanic antrum (cavity)
8. Auditory (Eustachian) tube
9. Auditory ossicles
10. Semicircular canals
11. Cochlea

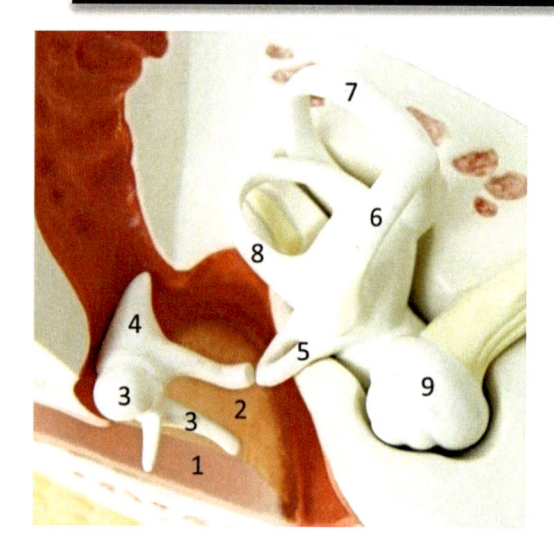

Figure 4-32. Auditory Ossicles

1. External acoustic meatus
2. Tympanic membrane
3. Malleus
4. Incus
5. Stapes
6. Anterior semicircular canal
7. Posterior semicircular canal
8. Lateral semicircular canal
9. Cochlea

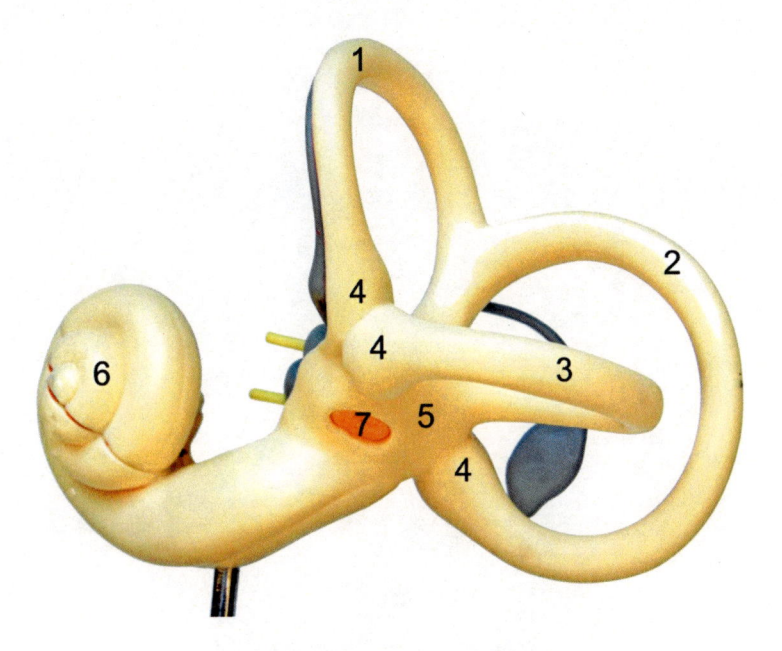

Figure 4-33. Inner Ear Labyrinth
(Copyright Somso)

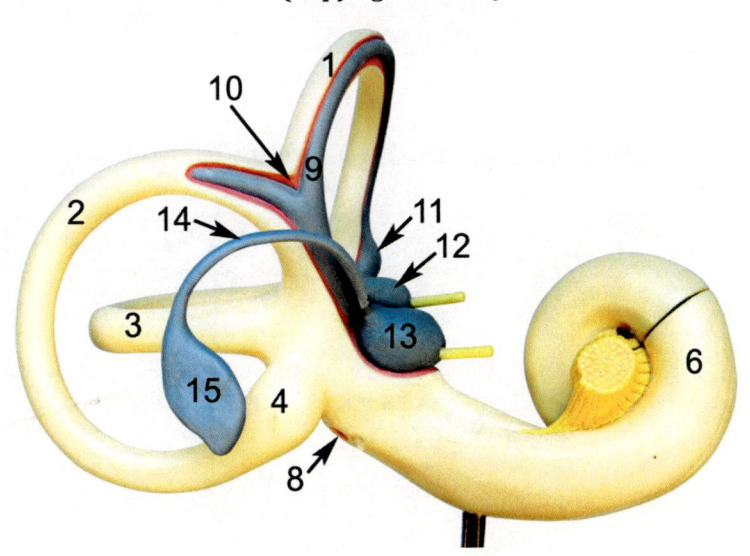

Figure 4-34. Inner Ear Labyrinth

1. Anterior semicircular canal
2. Posterior semicircular canal
3. Lateral semicircular canal
4. Ampulla of semicircular canal
5. Vestibule
6. Cochlea
7. Oval window
8. Round window
9. Semicircular duct
10. Perilymph (pink)
11. Ampulla of semicircular duct
12. Utricle
13. Saccule
14. Endolymphatic duct
15. Endolymphatic sac

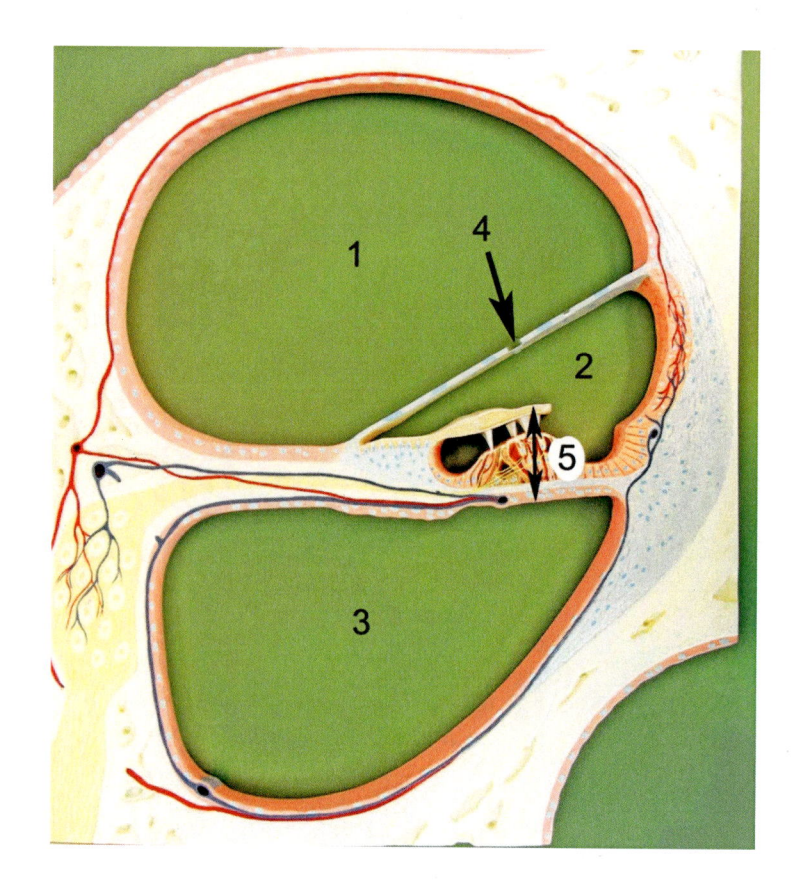

Figure 4-35. Cochlea – cross section

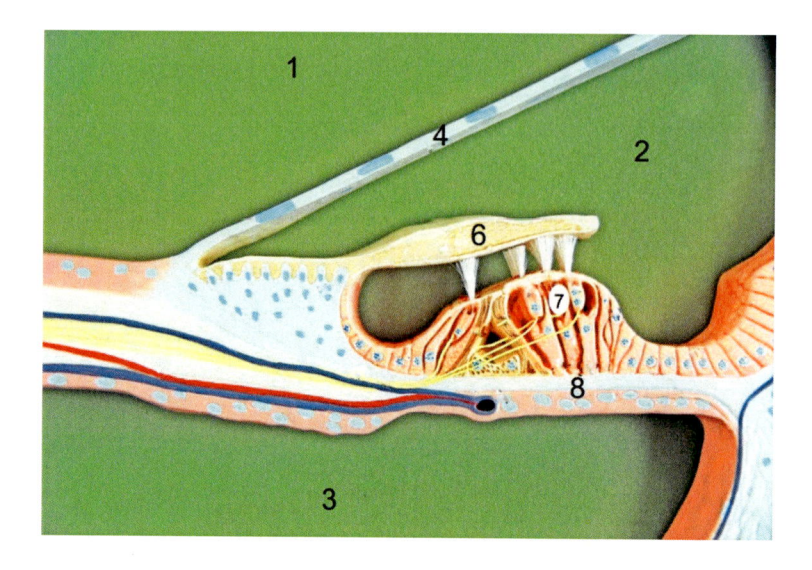

Figure 4-36. Cochlea Organ of Corti – cross section
(Copyright Somso)

1. Vestibular duct containing perilymph
2. Cochlear duct containing endolymph
3. Tympanic duct containing perilymph
4. Vestibular membrane
5. Organ of Corti (spiral organ)
6. Tectorial membrane
7. Hair cells
8. Basilar membrane

Chapter 5
Cardiovascular System Models

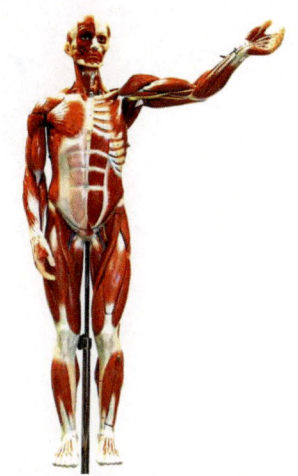

Male Muscle Figure
(Copyright Somso)

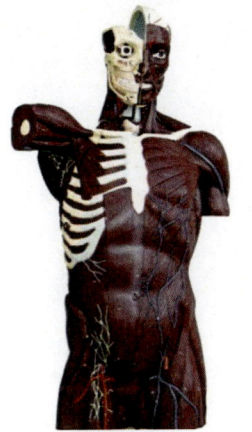

Torso
(Copyright Somso)

Torso
(Copyright 3B Scientific
)

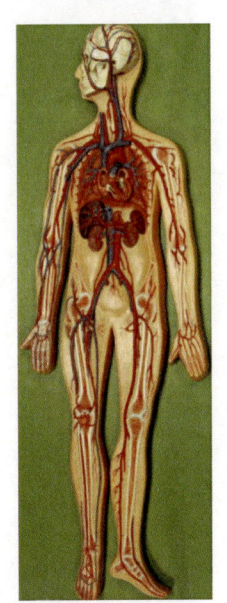

Circulatory Model – Flat
(Copyright Somso)

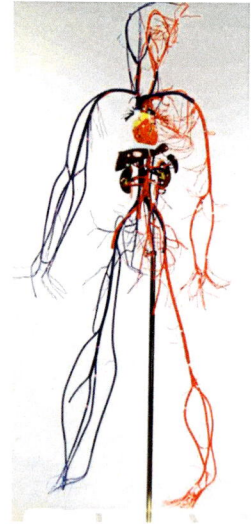

Circulatory Model – Wire
(purchased from Wards Scientific)

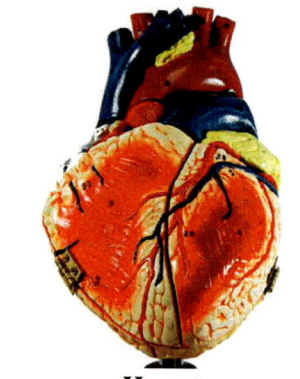

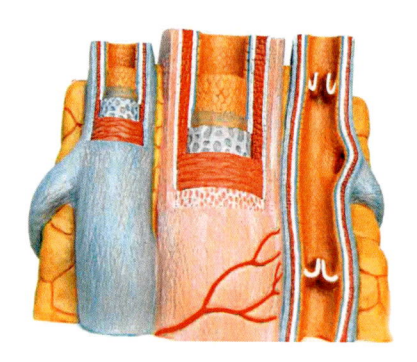

Blood Vessels
(Copyright 3B Scientific)

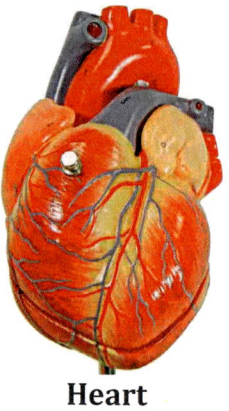

Heart
(Copyright Laerdal Medical)

Heart
(Copyright Somso)

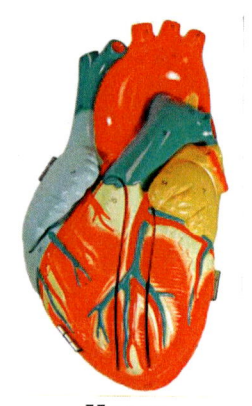

Heart
(Copyright Denoyer-Geppert)

Male Muscle Figure

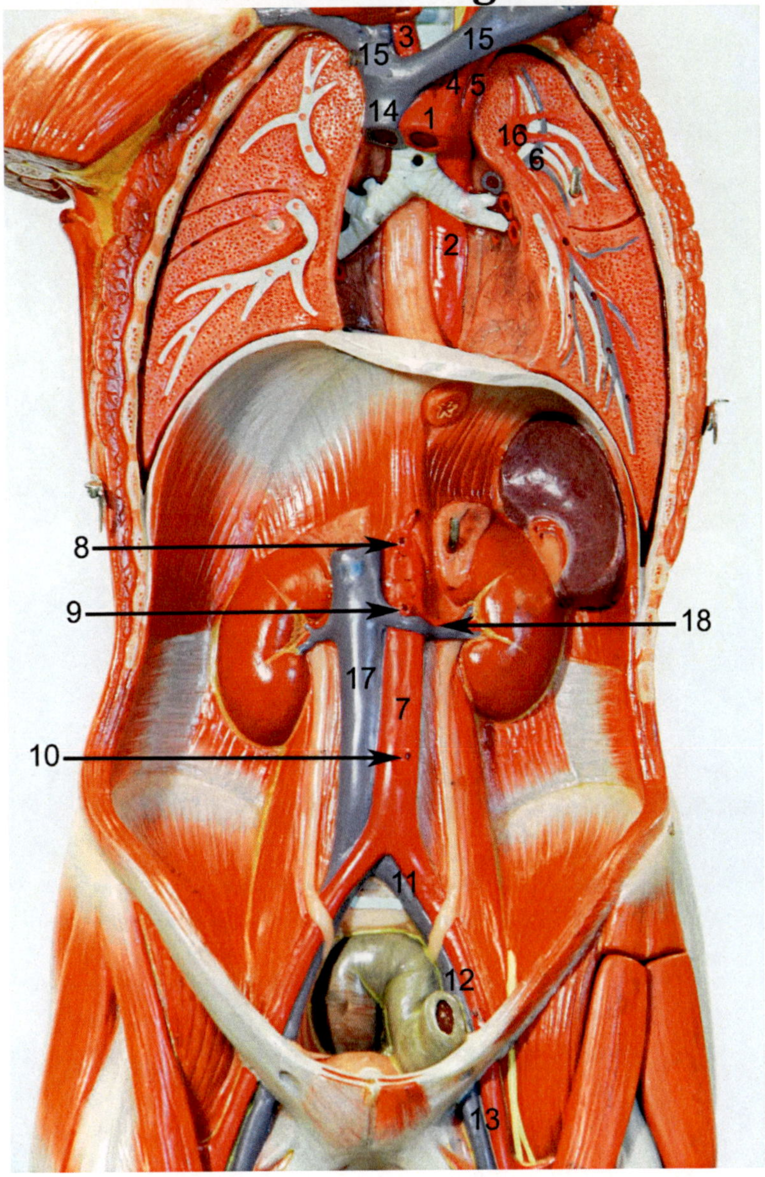

Figure 5-1. Vessels - ventral cavity
(Copyright Somso)

1. Ascending aorta
2. Thoracic aorta
3. Brachiocephalic artery
4. Left common carotid
5. Left subclavian artery
6. Pulmonary artery
7. Abdominal aorta
8. Celiac artery (trunk)
9. Superior mesenteric artery
10. Inferior mesenteric artery
11. Common iliac artery and vein
12. External iliac artery and vein
13. Femoral artery and vein
14. Superior vena cava
15. Brachiocephalic vein
16. Pulmonary vein
17. Inferior vena cava
18. Renal artery and vein

Torso – 3B

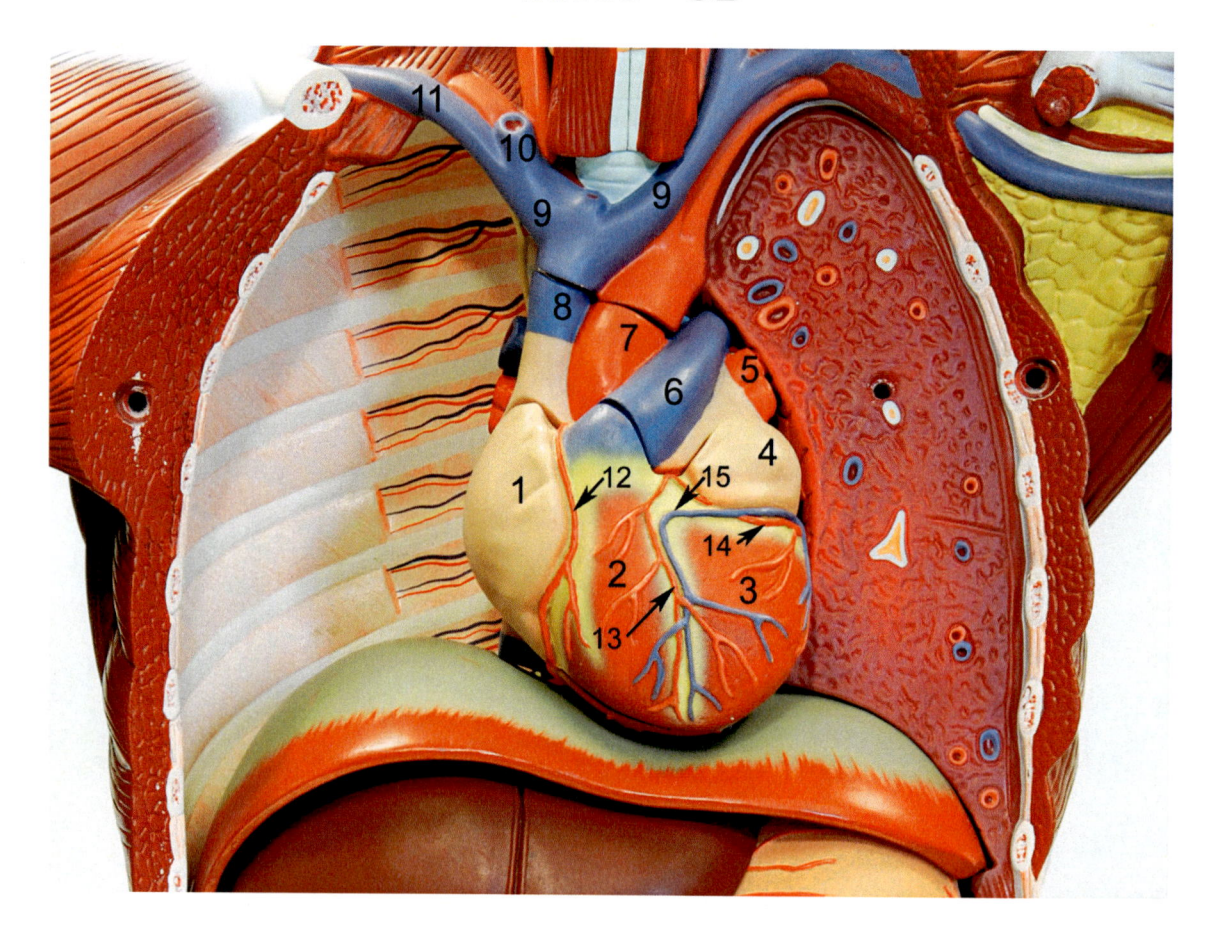

Figure 5-2. Heart and Vessels
(Copyright 3B Scientific)

1. Right atrium
2. Right ventricle
3. Left ventricle
4. Left atrium
5. Pulmonary veins
6. Pulmonary trunk (artery)
7. Ascending aorta
8. Superior vena cava
9. Brachiocephalic vein
10. Internal jugular vein
11. Subclavian vein
12. Right coronary artery
13. Anterior interventricular branch (left anterior descending)
14. Circumflex branch
15. Great cardiac vein

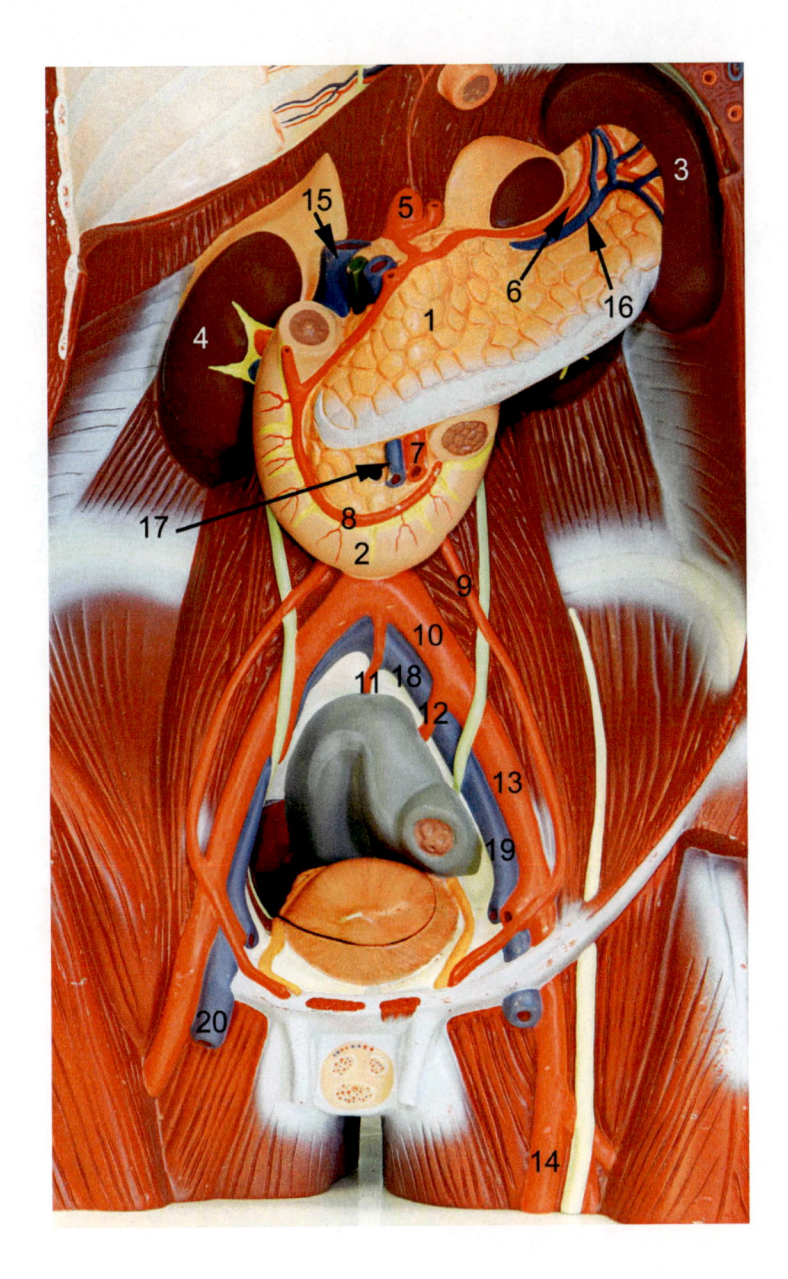

Figure 5-3. Vessels – abdominopelvic cavity
(Copyright 3B Scientific)

1. Pancreas
2. Duodenum
3. Spleen
4. Kidney
5. Abdominal aorta
6. Splenic artery
7. Superior mesenteric artery
8. Pancreaticoduodenal artery
9. Gonadal artery
10. Common iliac artery
11. Median sacral artery
12. Internal iliac artery
13. External iliac artery
14. Femoral artery
15. Inferior vena cava
16. Splenic vein
17. Superior mesenteric vein
18. Common iliac vein
19. External iliac vein
20. Femoral vein

Torso – Somso

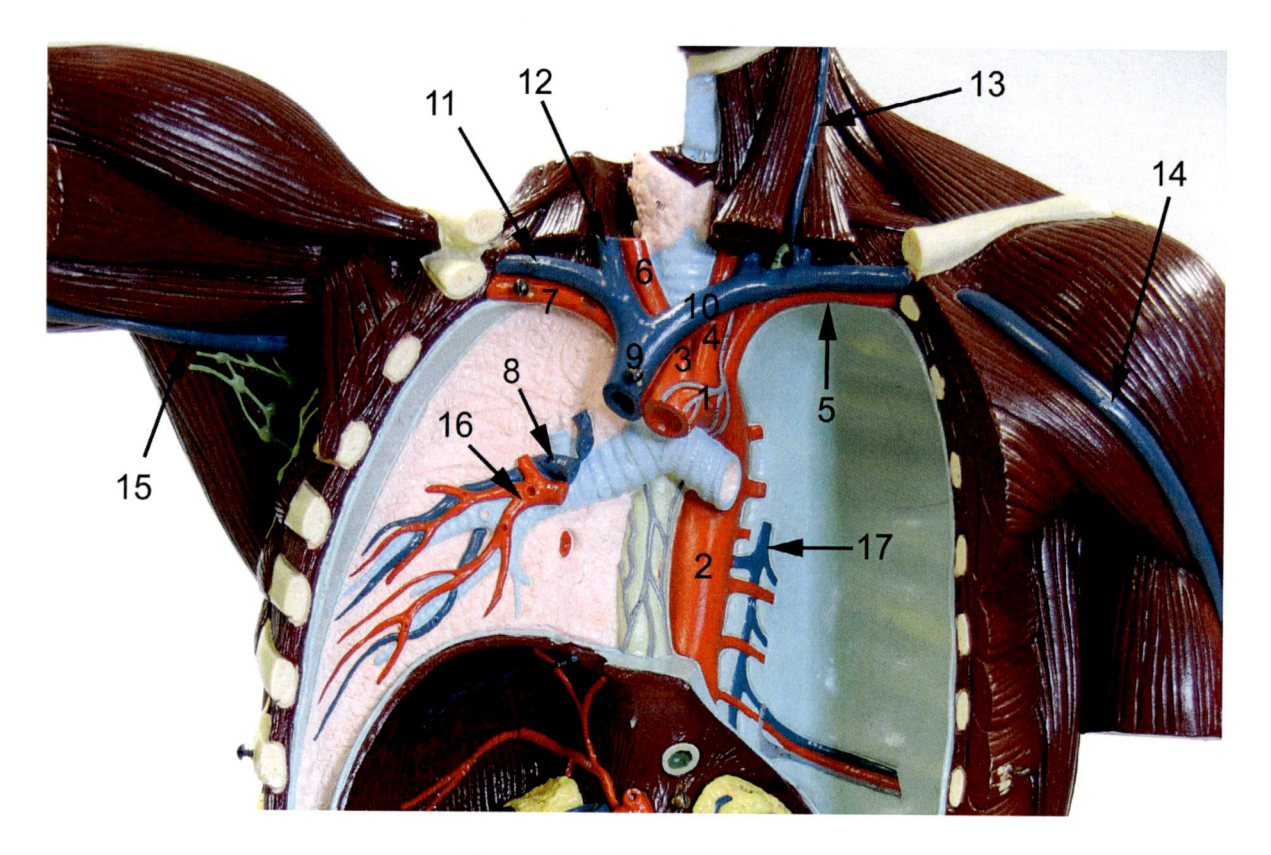

Figure 5-4. Vessels - thorax
(Copyright Somso)

1. Aortic arch
2. Thoracic aorta
3. Brachiocephalic trunk (artery)
4. Left common carotid
5. Left subclavian artery
6. Right common carotid
7. Right subclavian artery
8. Pulmonary artery
9. Superior vena cava
10. Brachiocephalic vein
11. Subclavian vein
12. Internal jugular vein
13. External jugular vein
14. Cephalic vein
15. Axillary vein
16. Pulmonary vein
17. Accessory hemiazygos

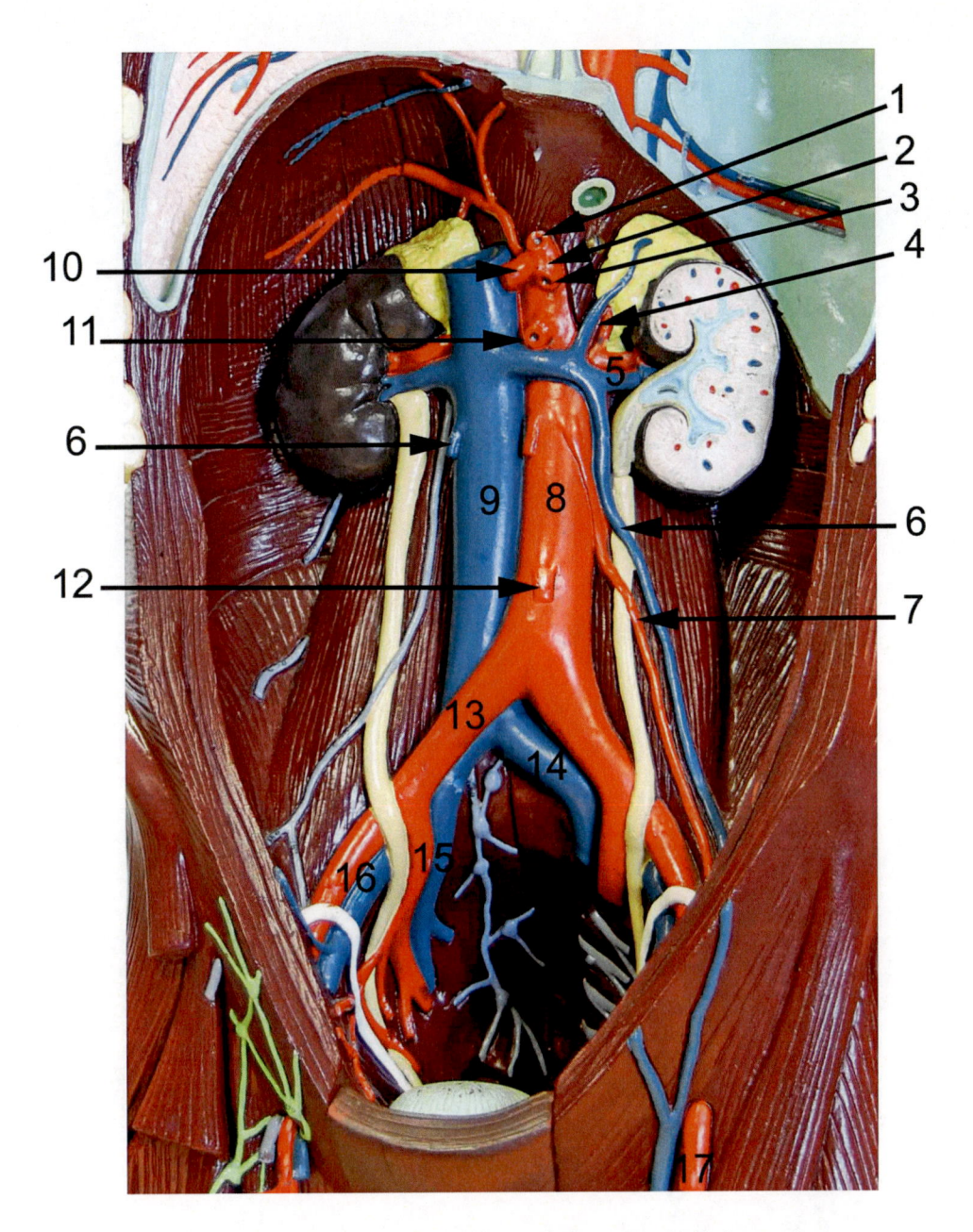

Figure 5-5. Vessels – abdominopelvic cavity
(Copyright Somso)

1. Left gastric artery
2. Celiac trunk (artery)
3. Splenic artery
4. Adrenal artery and vein
5. Renal artery and vein
6. Gonadal vein
7. Gonadal artery
8. Abdominal aorta
9. Inferior vena cava
10. Hepatic artery
11. Superior mesenteric artery
12. Inferior mesenteric artery
13. Common iliac artery
14. Common iliac vein
15. Internal iliac artery and vein
16. External iliac artery and vein
17. Femoral artery

Flat Circulatory Model

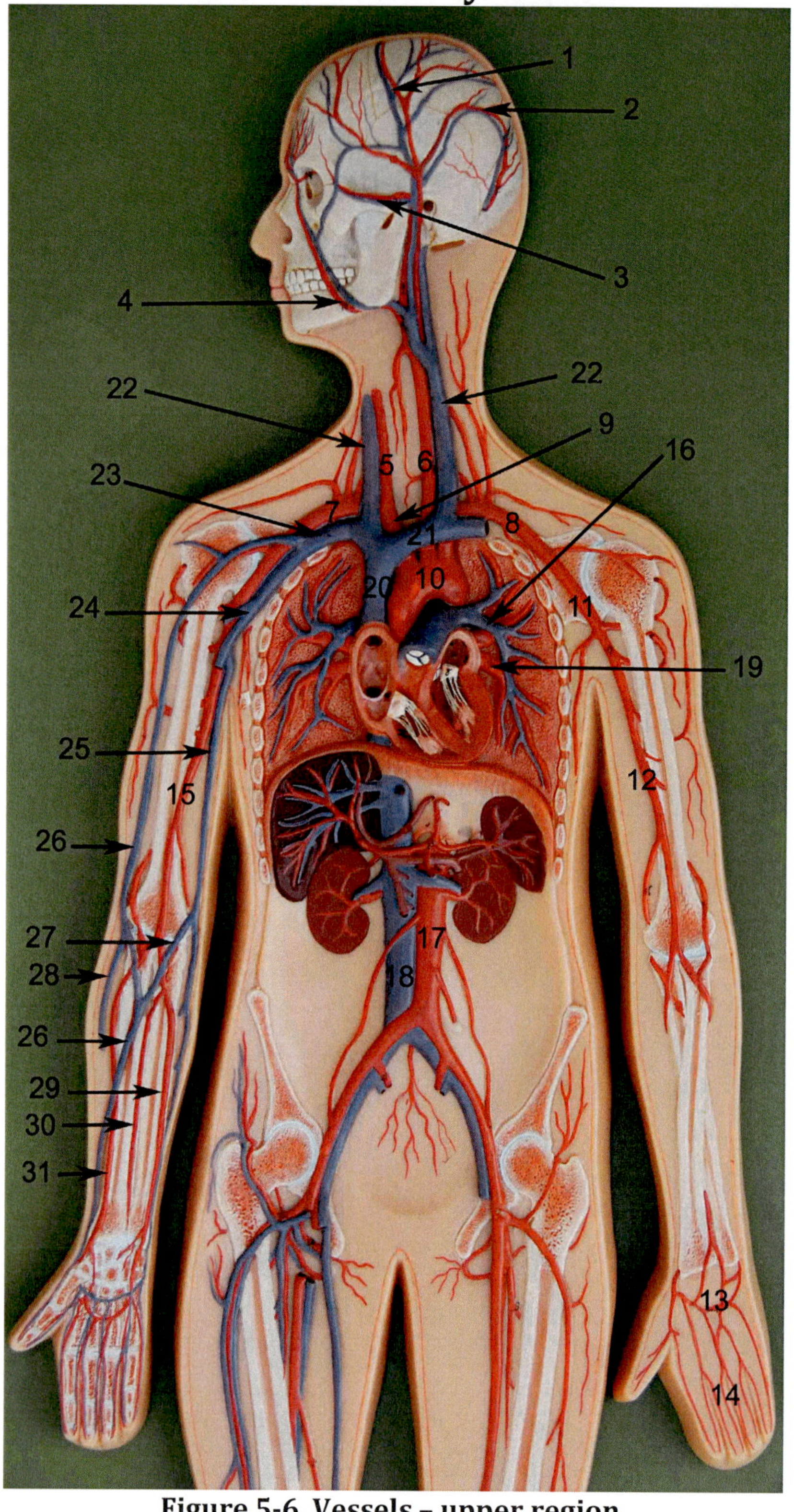

Figure 5-6. Vessels – upper region
(Copyright Somso)

Identification Key for Figure 5-6

1. Temporal vessels
2. Parietal vessels
3. Ophthalmic vessels
4. Facial vessels
5. Right common carotid
6. Left common carotid
7. Right subclavian artery
8. Left subclavian artery
9. Brachiocephalic trunk (artery)
10. Aortic arch
11. Axillary artery
12. Brachial artery
13. Palmar arches
14. Digital arteries
15. Brachial artery
16. Pulmonary artery
17. Abdominal aorta
18. Inferior vena cava
19. Pulmonary vein
20. Superior vena cava
21. Brachiocephalic vein
22. Internal jugular vein
23. Subclavian vein
24. Axillary vein
25. Basilic vein
26. Cephalic vein
27. Median cubital vein
28. Accessory cephalic vein
29. Ulnar artery
30. Anterior interosseous artery
31. Radial artery

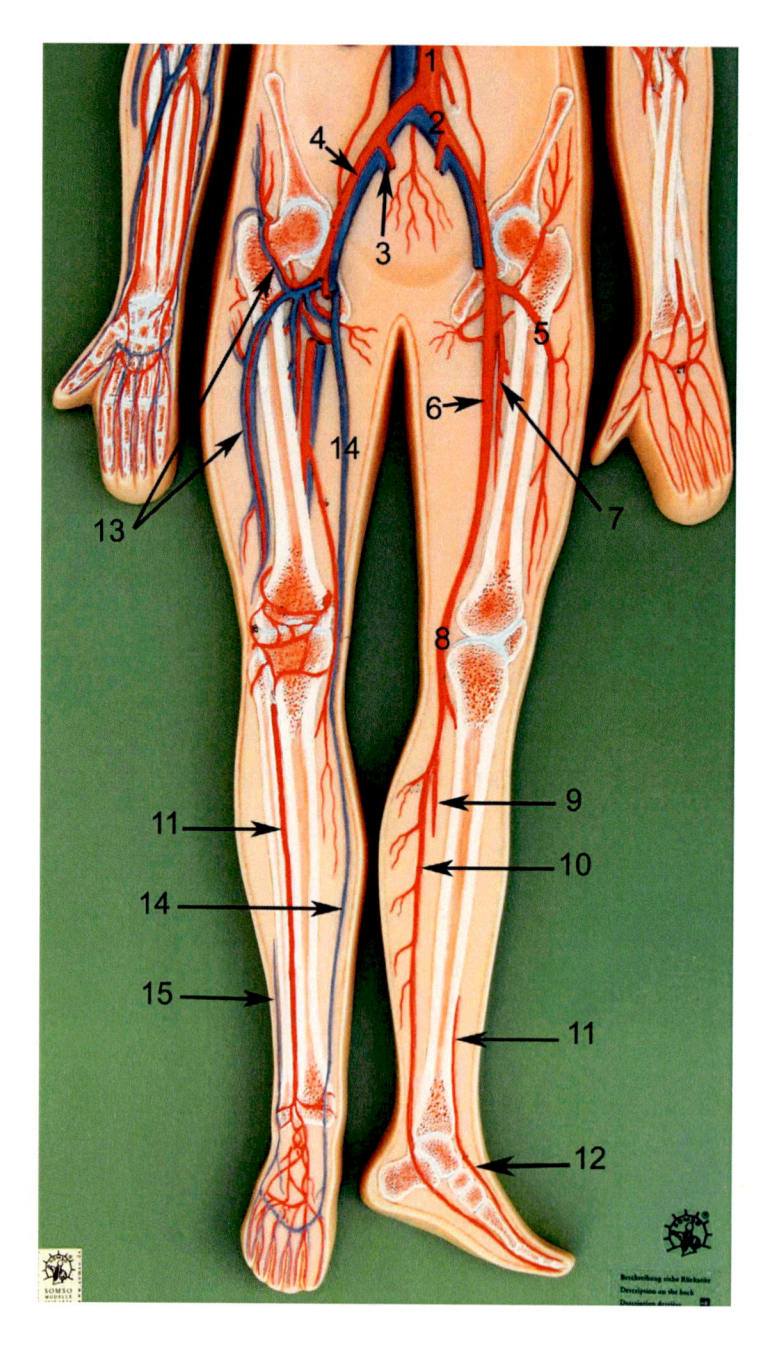

Figure 5-7. Vessels – lower region
(Copyright Somso)

1. Abdominal aorta
2. Common iliac artery and vein
3. Internal iliac artery and vein
4. External iliac artery and vein
5. Lateral femoral circumflex artery
6. Femoral artery
7. Deep femoral artery
8. Popliteal artery
9. Fibular (peroneal) artery
10. Posterior tibial artery
11. Anterior tibial artery
12. Dorsalis pedis artery
13. Lateral femoral circumflex veins
14. Great saphenous vein
15. Small saphenous vein

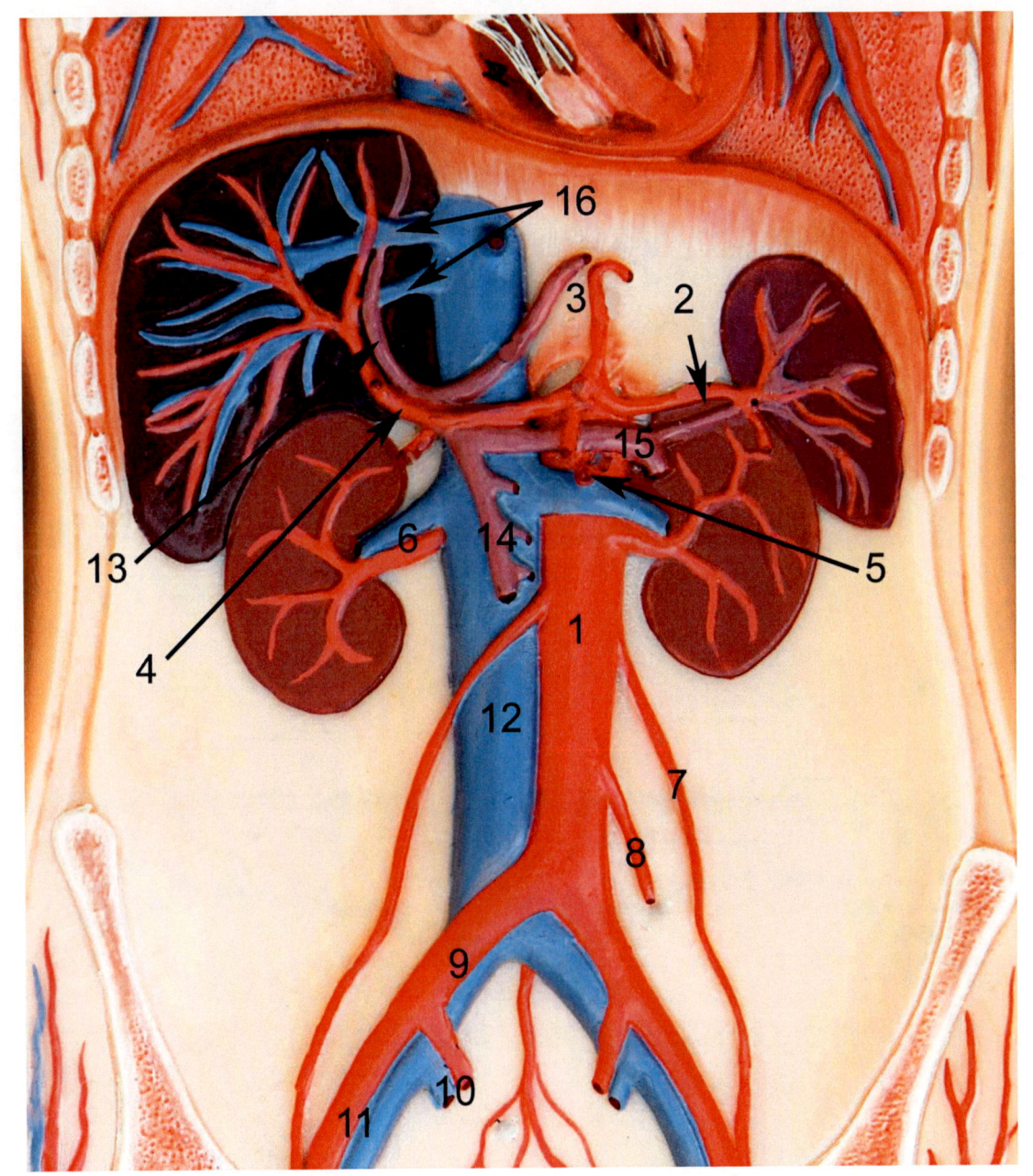

Figure 5-8. Vessels - abdominopelvic cavity
(Copyright Somso)

1. Abdominal aorta
2. Splenic artery and vein
3. Left gastric artery and vein
4. Hepatic artery
5. Superior mesenteric artery
6. Renal artery and vein
7. Gonadal artery
8. Inferior mesenteric artery
9. Common iliac artery and vein
10. Internal iliac artery and vein
11. External iliac artery and vein
12. Inferior vena cava
13. Hepatic portal vein
14. Superior mesenteric vein
15. Inferior mesenteric vein
16. Hepatic veins

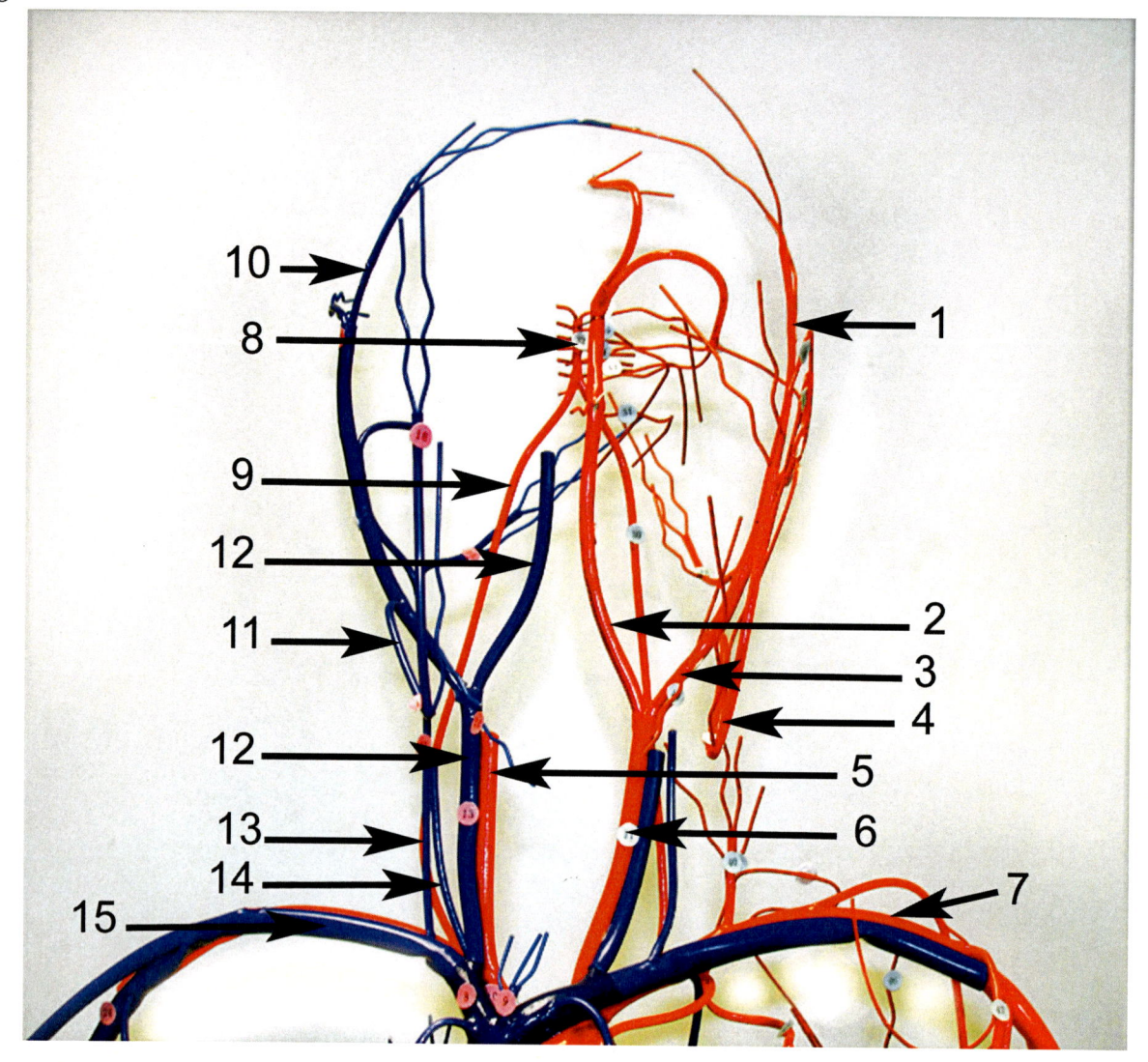

Figure 5-9. Vessels – head and neck

1. Superficial temporal artery
2. Internal carotid
3. External carotid
4. Facial artery
5. Right common carotid
6. Left common carotid
7. Left subclavian artery
8. Basilar artery
9. Vertebral artery
10. Superficial temporal vein
11. Facial vein
12. Internal jugular vein
13. External jugular vein
14. Vertebral vein
15. Subclavian vein

99

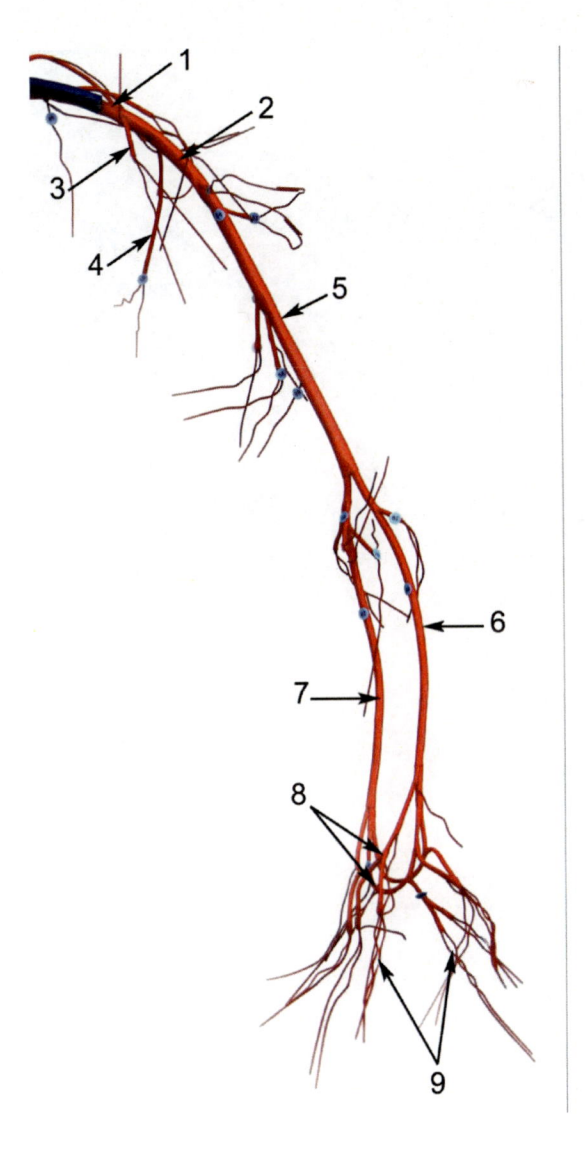

Figure 5-10. Arteries – left "arm"

1. Subclavian artery
2. Axillary artery
3. Lateral thoracic artery
4. Subscapular artery
5. Brachial artery
6. Radial artery
7. Ulnar artery
8. Palmar arches
9. Digital arteries

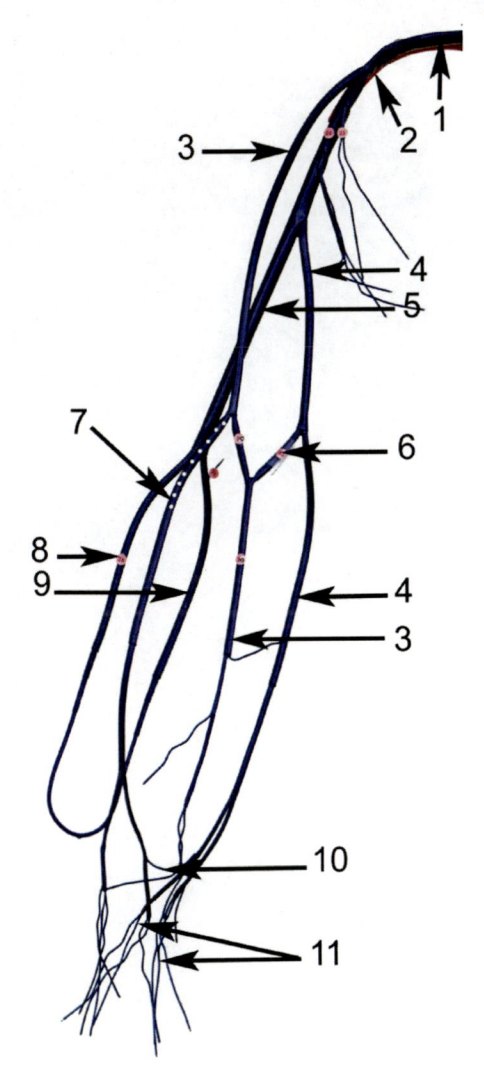

Figure 5-11. Veins – right "arm"

1. Subclavian vein
2. Axillary vein
3. Cephalic vein
4. Basilic vein
5. Brachial vein
6. Median cubital vein
7. Accessory cephalic vein
8. Radial vein
9. Ulnar vein
10. Palmar venous arches
11. Digital veins

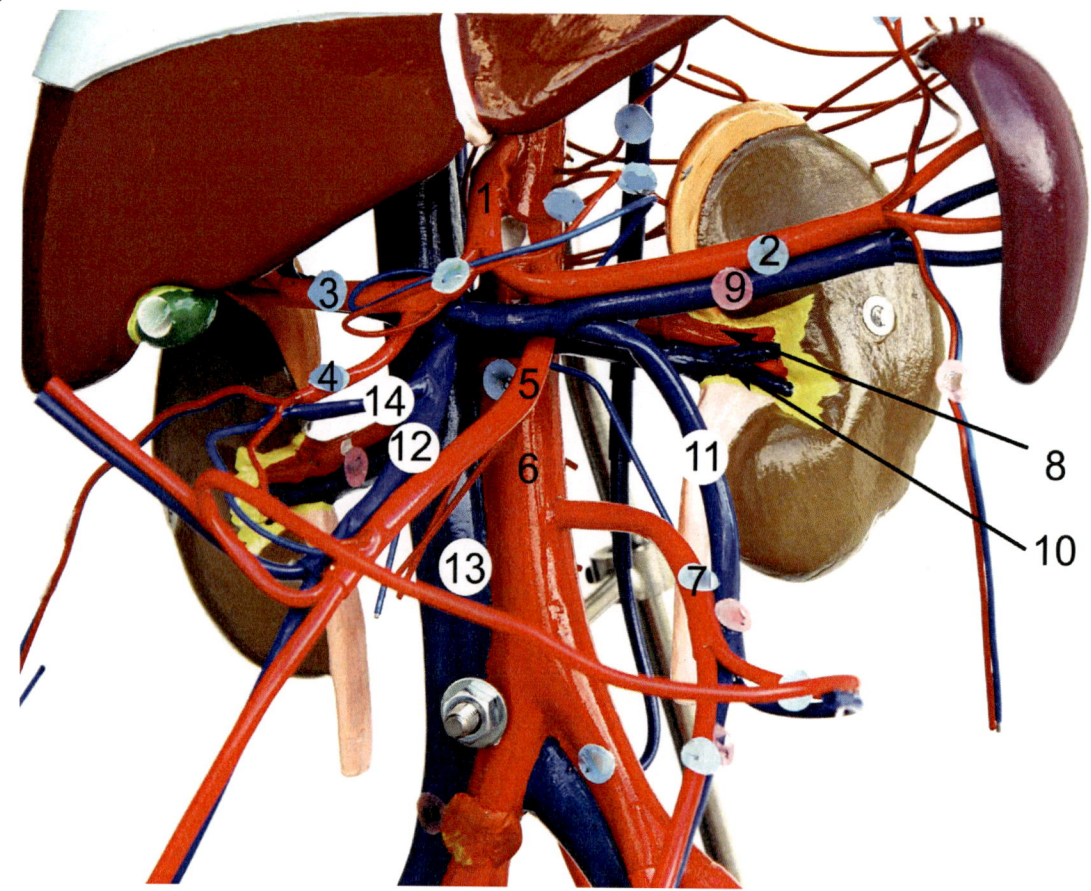

Figure 5-12. Vessels – anterior abdominal region

1. Celiac trunk (artery)
2. Splenic artery
3. Hepatic artery
4. Gastroduodenal artery
5. Superior mesenteric artery
6. Abdominal aorta
7. Inferior mesenteric artery
8. Renal artery
9. Splenic vein
10. Renal vein
11. Inferior mesenteric vein
12. Superior mesenteric vein
13. Inferior vena cava
14. Gastroduodenal vein

101

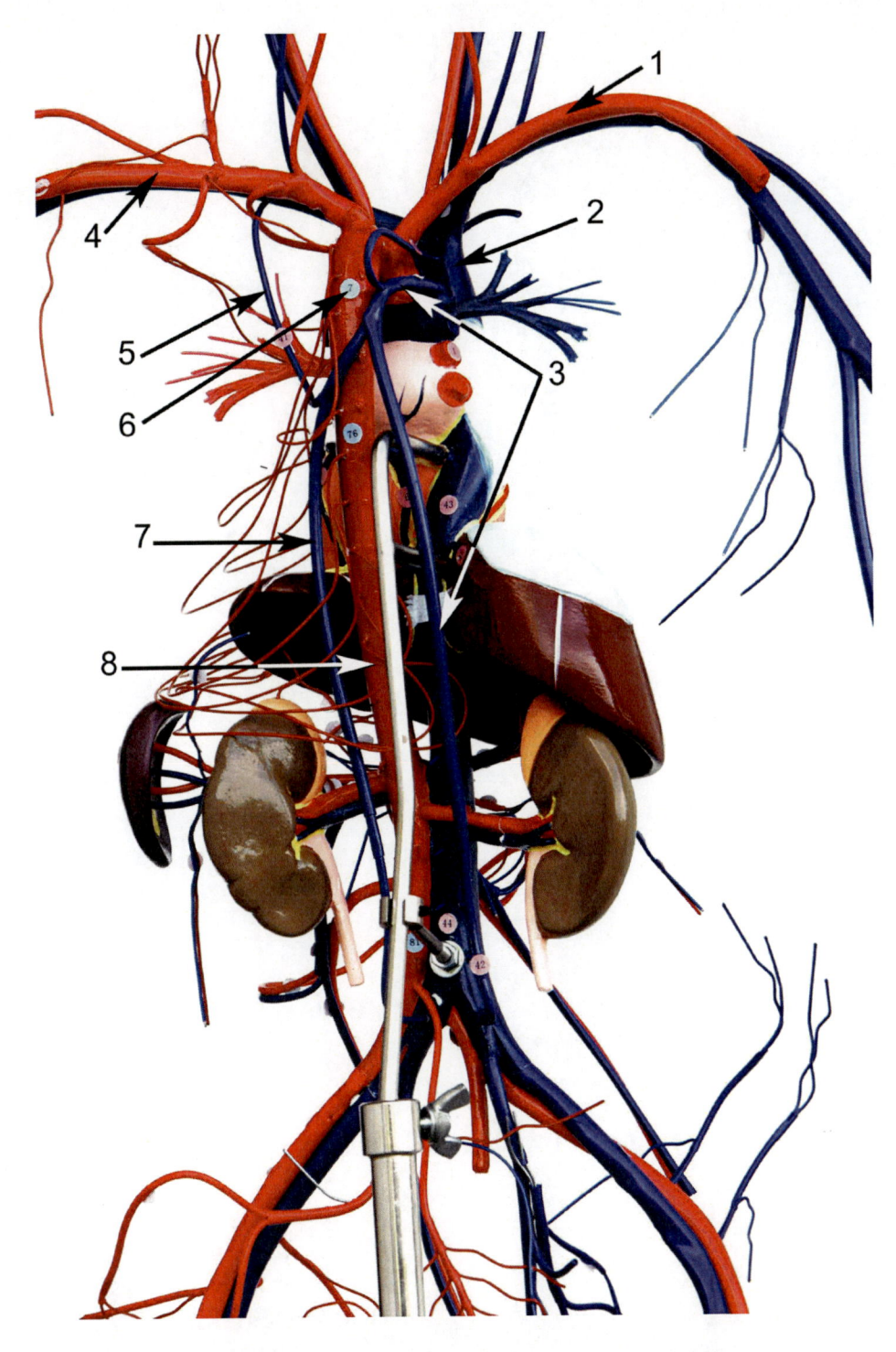

Figure 5-13. Vessels - posterior trunk

1. Right subclavian artery
2. Superior vena cava
3. Azygos vein
4. Left subclavian artery
5. Accessory hemiazygos vein
6. Thoracic aorta
7. Hemiazygos vein
8. Abdominal aorta

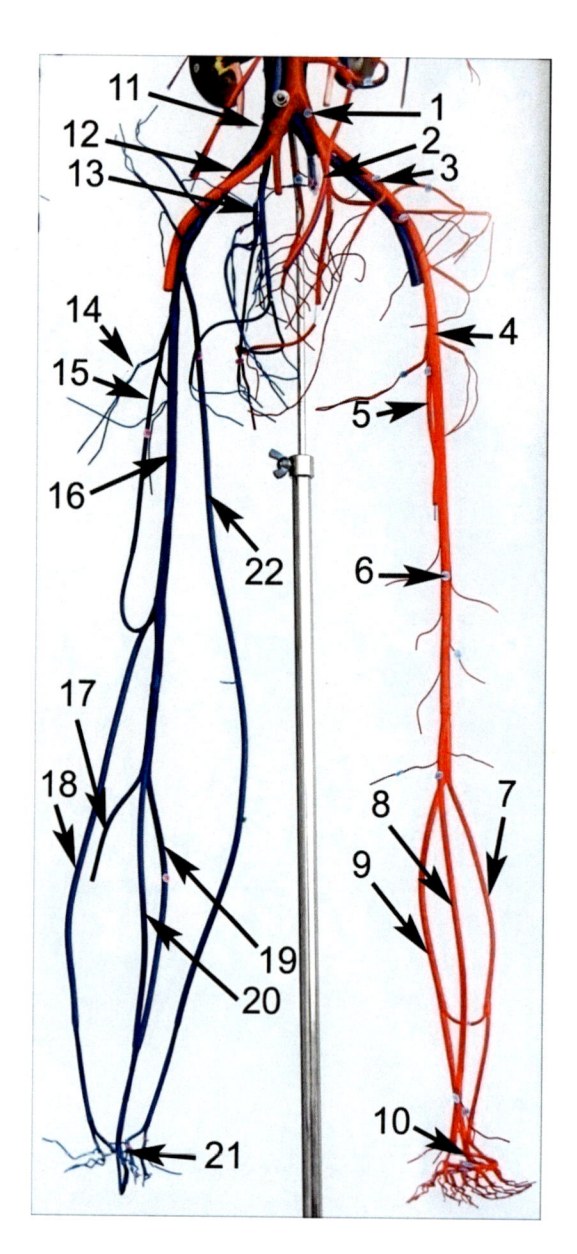

Figure 5-14. Vessels – lower extremities

1. Common iliac artery
2. Internal iliac artery
3. External iliac artery
4. Femoral artery
5. Deep femoral artery
6. Popliteal artery
7. Fibular (peroneal) artery
8. Anterior tibial artery
9. Posterior tibial artery
10. Dorsalis pedis artery
11. Common iliac vein
12. External iliac vein
13. Internal iliac vein
14. Femoral circumflex vein
15. Deep femoral vein
16. Femoral vein
17. Fibular (peroneal) vein
18. Small saphenous vein
19. Posterior tibial vein
20. Anterior tibial vein
21. Dorsal venous arch
22. Great saphenous vein

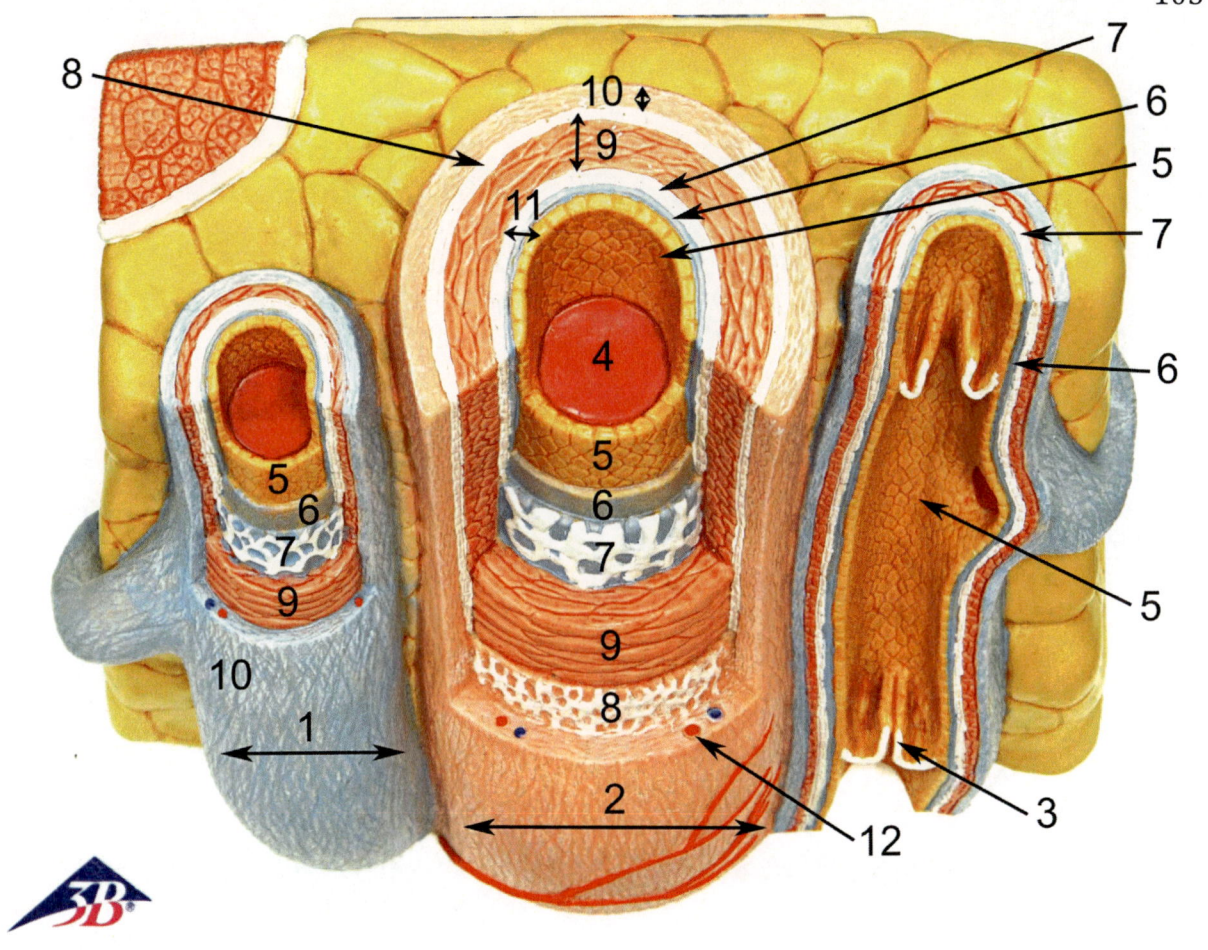

Figure 5-15. Blood Vessels (histology)
(Copyright 3B Scientific)

1. Vein
2. Artery
3. Valve in vein
4. Blood
5. Endothelium of tunica interna
6. Subendothelial connective tissue of tunica interna
7. Internal elastic membrane of tunica interna
8. External elastic membrane of tunica media
9. Tunica media
10. Tunica externa
11. Tunica interna
12. Vasa vasorum

1. Right atrium
2. Right ventricle
3. Left ventricle
4. Left atrium
5. Auricle of left atrium
6. Right coronary artery (RCA)
7. Left coronary artery (LCA)
8. Anterior interventricular branch (left anterior descending) of LCA
9. Circumflex branch of LCA
10. Great cardiac vein
11. Pulmonary trunk
12. Right pulmonary artery
13. Left pulmonary artery
14. Ligamentum arteriosum
15. Ascending aorta
16. Aortic arch
17. Superior vena cava
18. Brachiocephalic vein
19. Brachiocephalic trunk (artery)
20. Left common carotid
21. Left subclavian artery
22. Pulmonary veins
23. Coronary sinus
24. Posterior cardiac vein
25. Middle cardiac vein
26. Posterior interventricular branch (posterior descending artery) of RCA
27. Small cardiac vein
28. Inferior vena cava

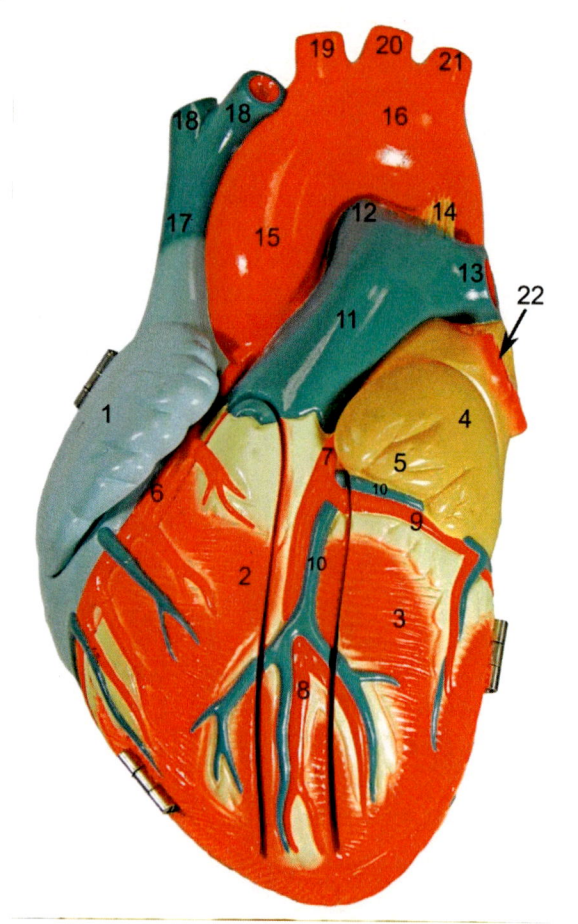

Figure 5-16. Heart - anterior
(Copyright Denoyer-Geppert)

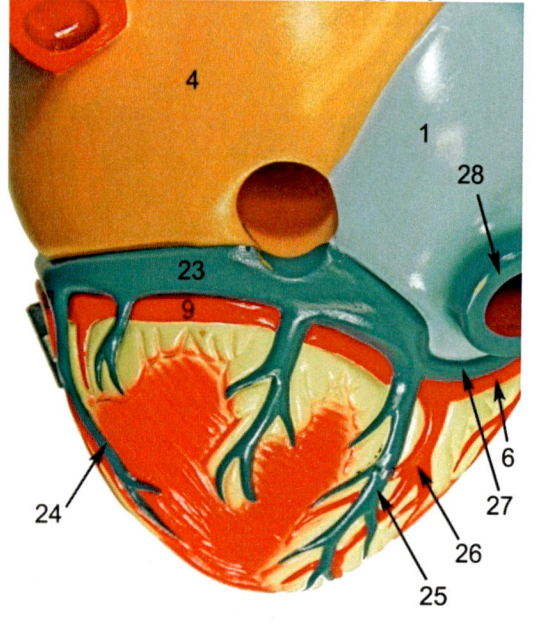

Figure 5-17. Heart - posterior

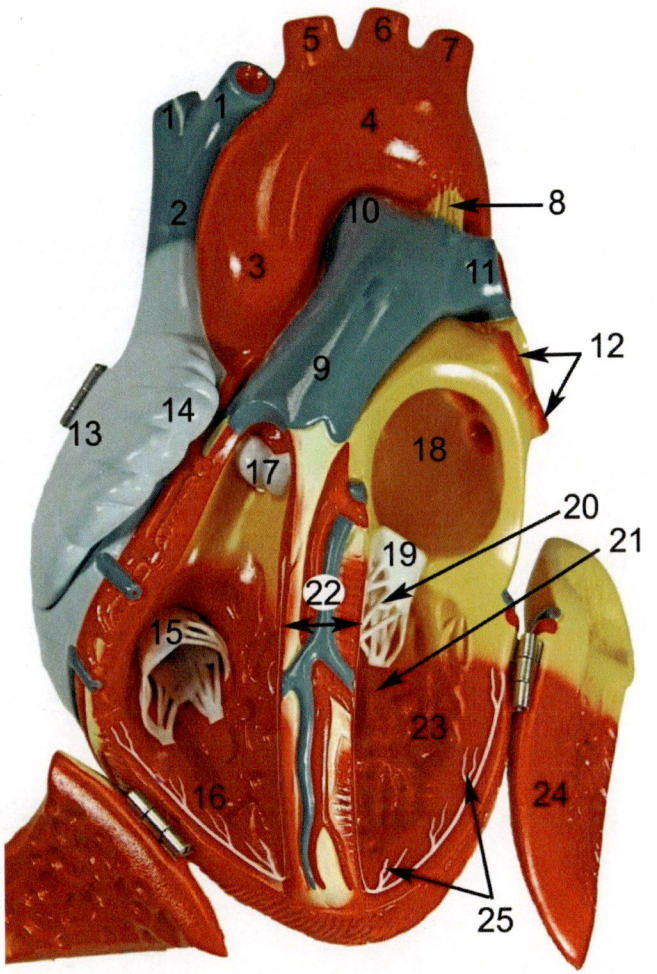

Figure 5-18. Heart - interior

1. Brachiocephalic veins
2. Superior vena cava
3. Ascending aorta
4. Aortic arch
5. Brachiocephalic trunk (artery)
6. Left common carotid
7. Left subclavian artery
8. Ligamentum arteriosum
9. Pulmonary trunk
10. Right pulmonary artery
11. Left pulmonary artery
12. Left pulmonary veins
13. Right atrium
14. Auricle of right atrium
15. Tricuspid valve
16. Right ventricle
17. Pulmonary valve
18. Left atrium
19. Mitral (bicuspid) valve
20. Chordae tendineae
21. Papillary muscle
22. Interventricular septum
23. Trabeculae carneae
24. Left ventricle (wall)
25. Subendocardial branches (Purkinje fibers)
26. Sinoatrial node
27. Atrioventricular node
28. Interatrial septum from right atrium
29. Right coronary artery
30. Right pulmonary vein

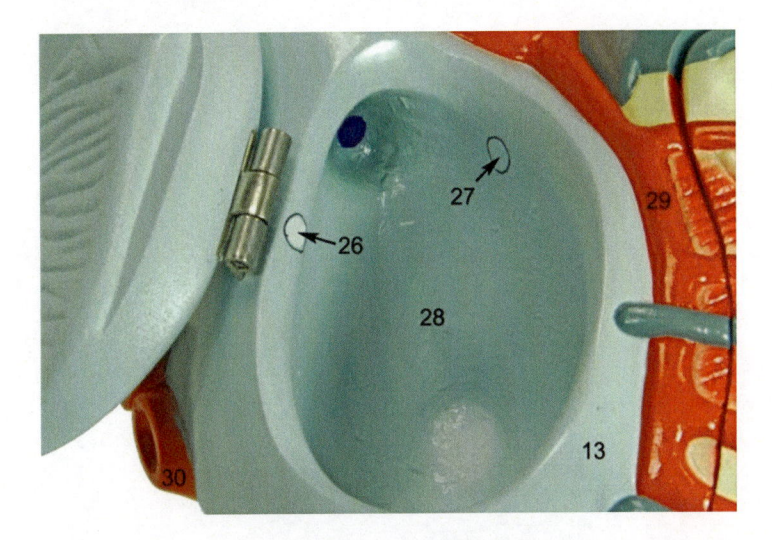

Figure 5-19. Heart – right atrium
(Copyright Denoyer-Geppert)

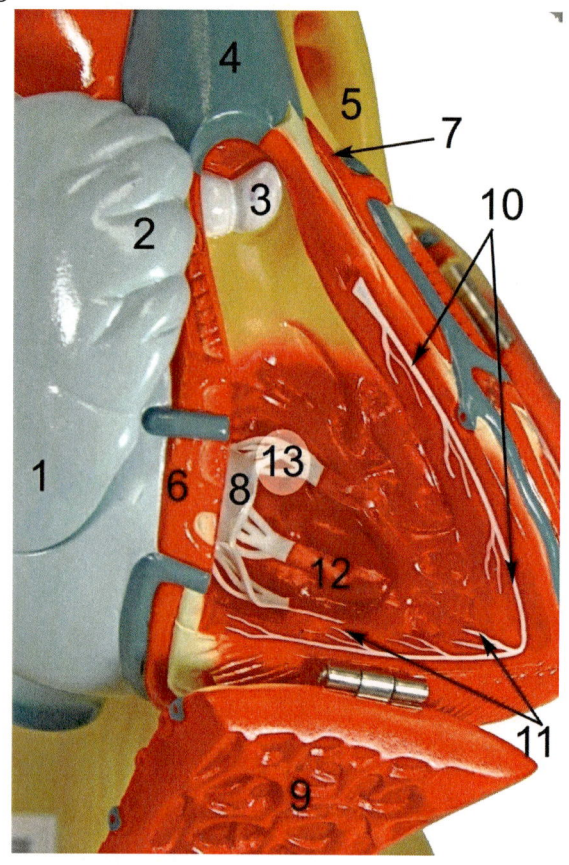

Figure 5-20. Heart – right ventricle
(Copyright Denoyer-Geppert)

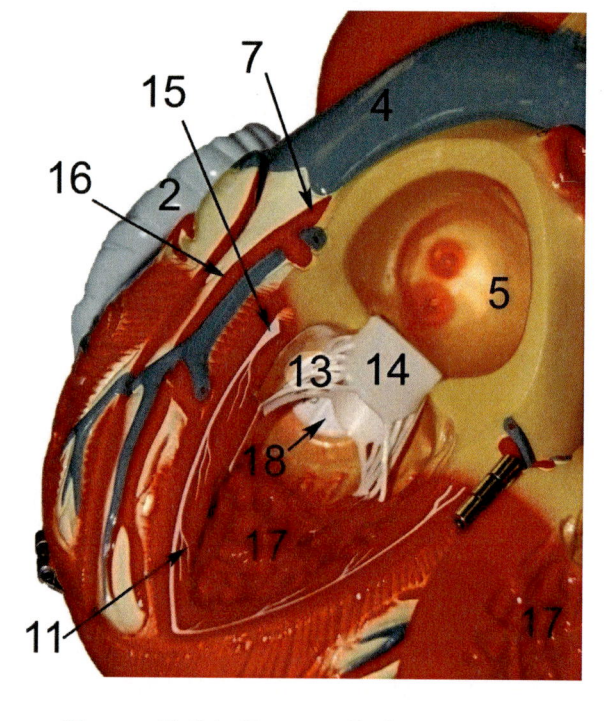

Figure 5-21. Heart – left ventricle

1. Right atrium
2. Auricle of right atrium
3. Pulmonary valve
4. Pulmonary trunk (artery)
5. Left atrium
6. Right coronary artery (RCA)
7. Left coronary artery (LCA)
8. Tricuspid valve
9. Right ventricle (wall)
10. Right bundle branch
11. Subendocardial branches (Purkinje fibers)
12. Papillary muscle
13. Chordae tendineae
14. Mitral (bicuspid) valve
15. Left bundle branch
16. Anterior interventricular branch of LCA
17. Trabeculae carneae of left ventricle
18. Aortic valve

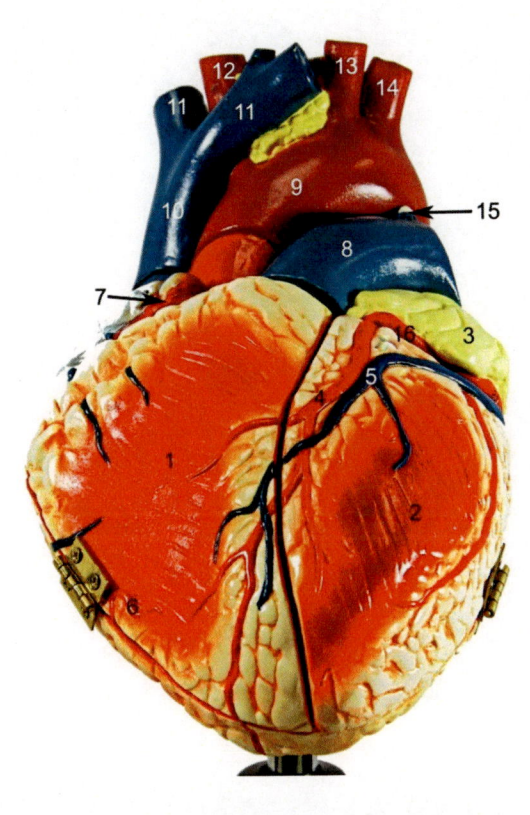

Figure 5-22. Heart - anterior
(Copyright Laerdal Medical)

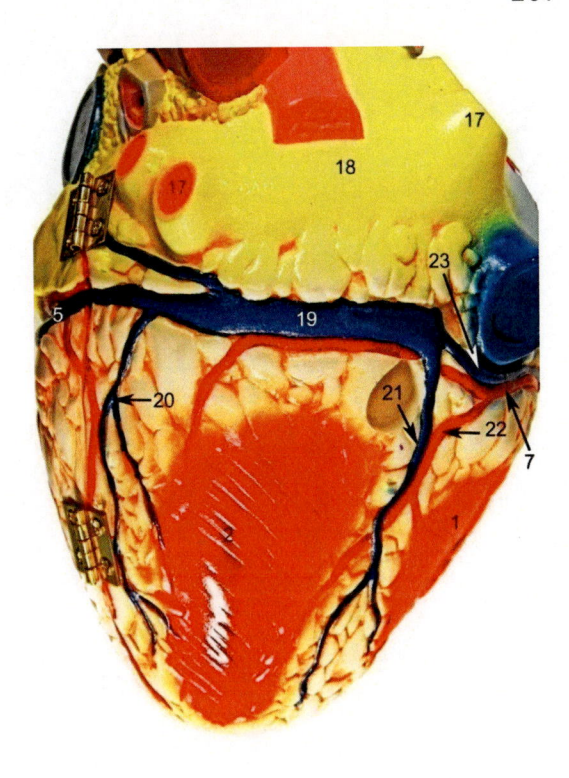

Figure 5-23. Heart - posterior

1. Right ventricle
2. Left ventricle
3. Auricle of left atrium
4. Anterior interventricular branch of left coronary artery (LCA)
5. Great cardiac vein
6. Marginal branch of RCA
7. Right coronary artery (RCA)
8. Pulmonary trunk
9. Ascending aorta
10. Superior vena cava
11. Brachiocephalic vein
12. Brachiocephalic trunk (artery)
13. Left common carotid
14. Left subclavian artery
15. Ligamentum arteriosum
16. Circumflex branch of LCA
17. Pulmonary veins
18. Left atrium
19. Coronary sinus
20. Posterior cardiac vein
21. Middle cardiac vein
22. Posterior interventricular branch of RCA
23. Small cardiac vein

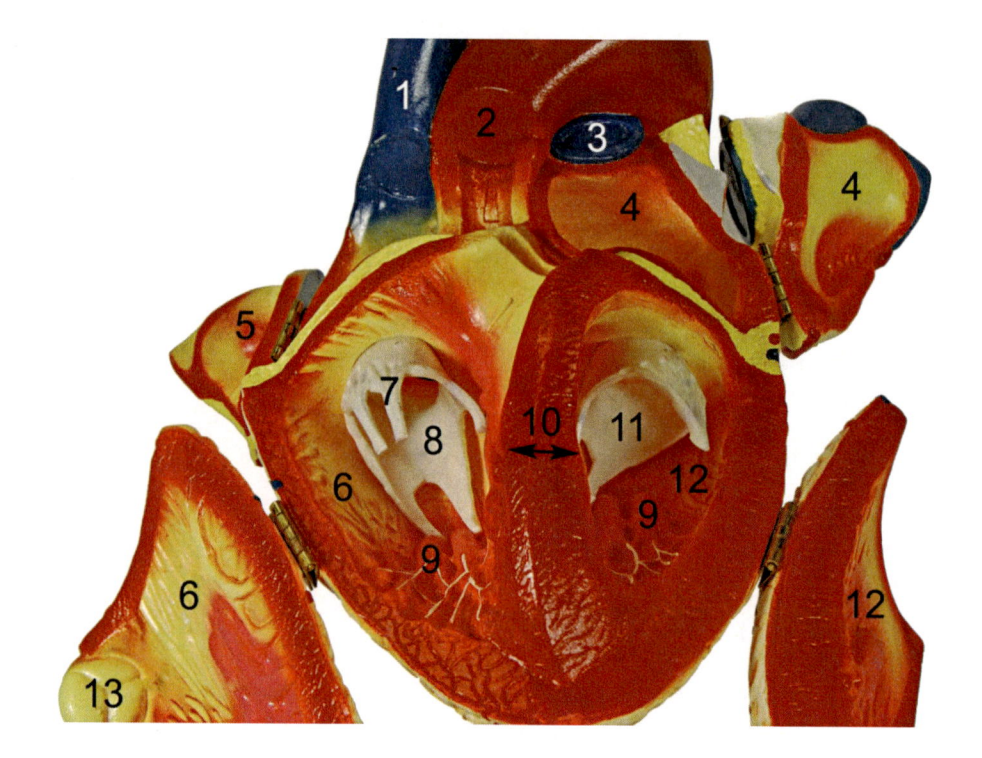

Figure 5-24. Heart - interior

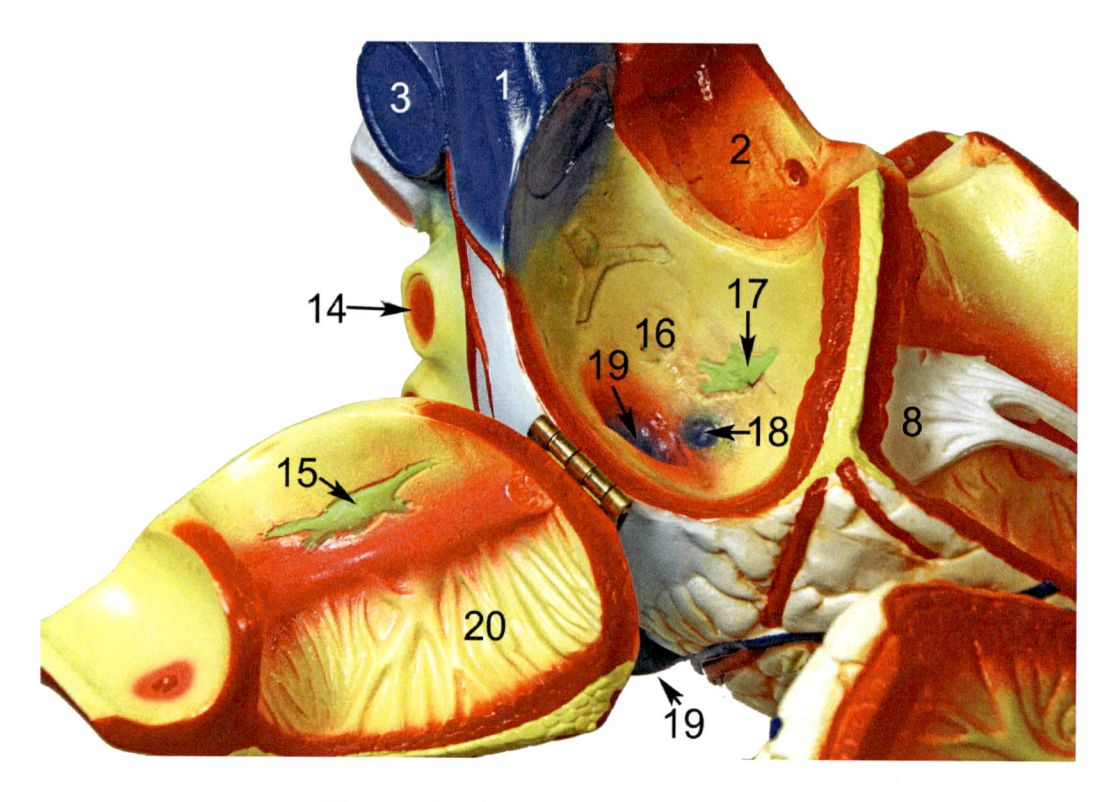

Figure 5-25. Heart – right atrium
(Copyright Laerdal Medical)

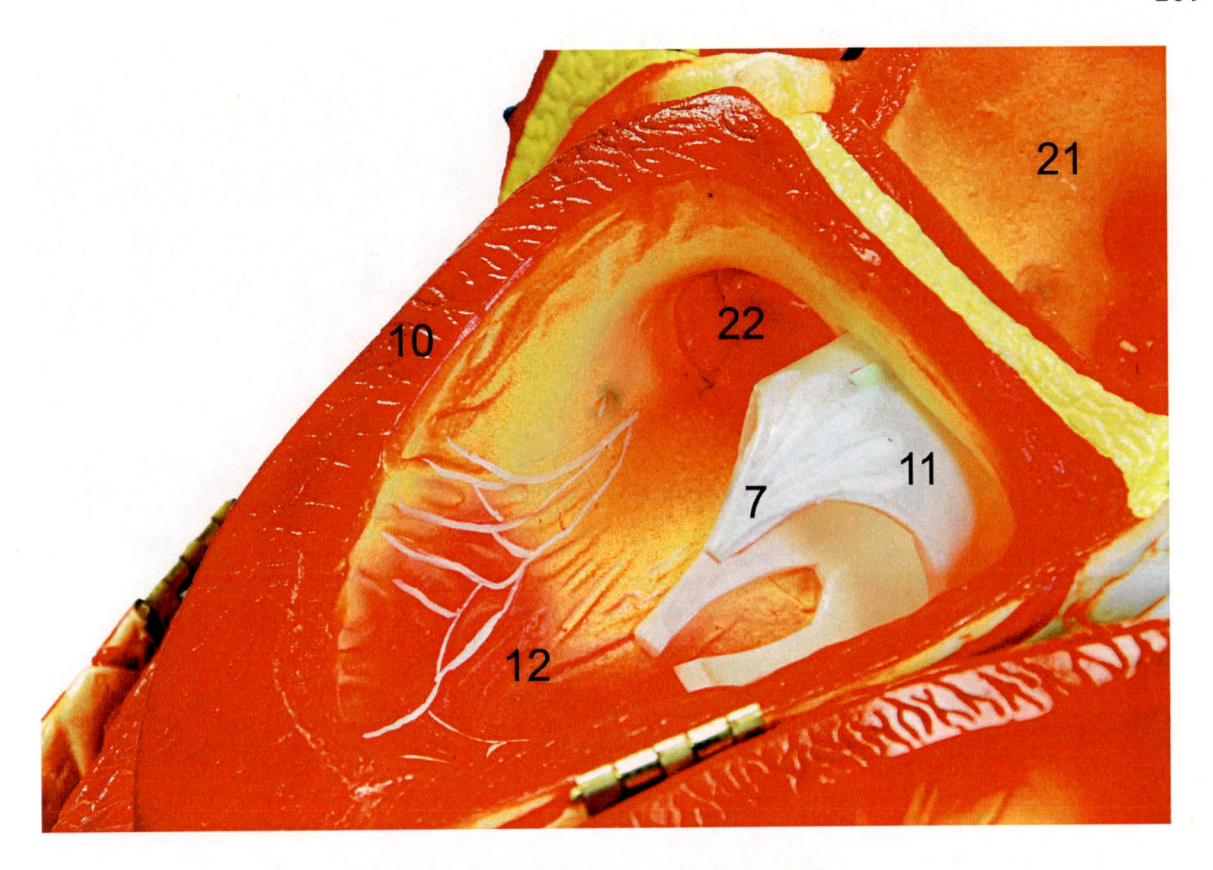

Figure 5-26. Heart – left ventricle
(Copyright Laerdal Medical)

Identification Key for figures 5-24 to 5-26

1. Superior vena cava
2. Aorta
3. Right pulmonary artery
4. Left atrium
5. Right atrium
6. Right ventricle
7. Chordae tendineae
8. Tricuspid valve
9. Papillary muscle
10. Interventricular septum
11. Mitral (bicuspid) valve
12. Left ventricle
13. Pulmonary valve
14. Right pulmonary vein
15. Sinoatrial node
16. Fossa ovalis
17. Atrioventricular node
18. Opening for coronary sinus
19. Inferior vena cava
20. Pectinate muscles
21. Interatrial septum from left atrium
22. Aortic valve

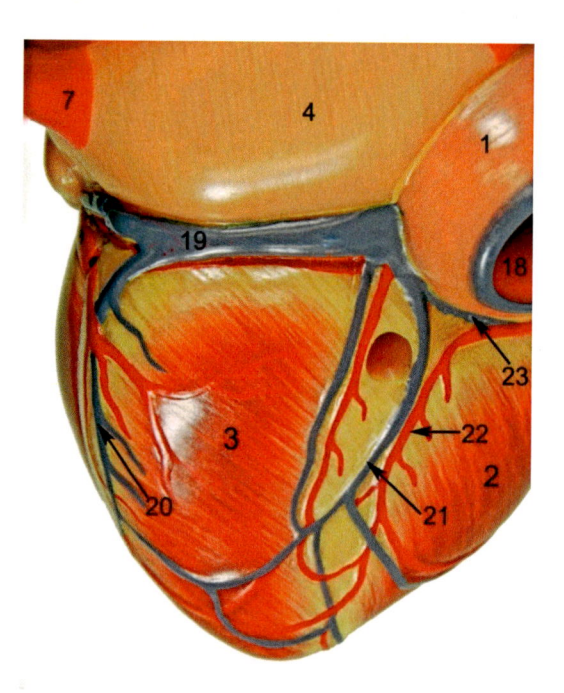

Figure 5-28. Heart – posterior
(Copyright Somso)

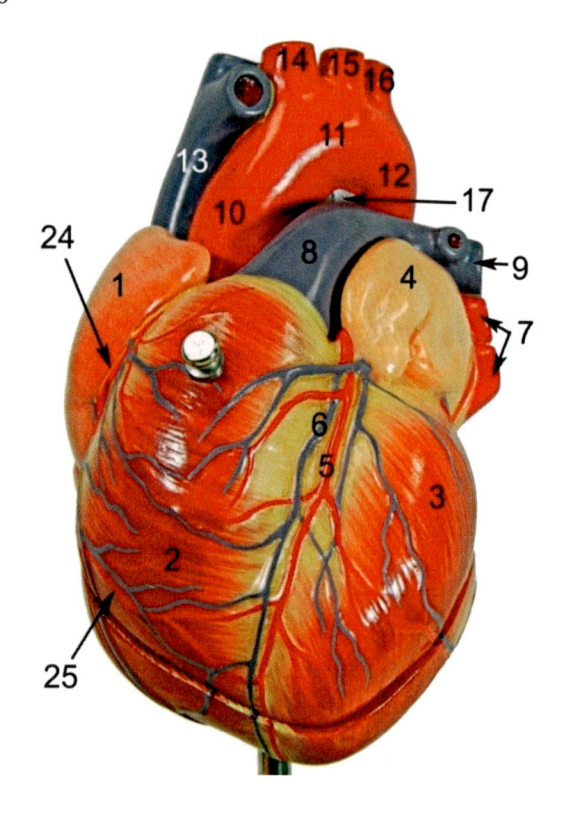

Figure 5-27. Heart – anterior

1. Right atrium
2. Right ventricle
3. Left ventricle
4. Left atrium
5. Anterior interventricular branch of LCA
6. Great cardiac vein
7. Left pulmonary veins
8. Pulmonary trunk
9. Left pulmonary artery
10. Ascending aorta
11. Aortic arch
12. Descending aorta
13. Superior vena cava
14. Brachiocephalic trunk (artery)
15. Left common carotid
16. Left subclavian artery
17. Ligamentum arteriosum
18. Inferior vena cava
19. Coronary sinus
20. Posterior cardiac vein
21. Middle cardiac vein
22. Posterior interventricular branch of RCA
23. Small cardiac vein
24. Right coronary artery (RCA)
25. Marginal branch of RCA

111

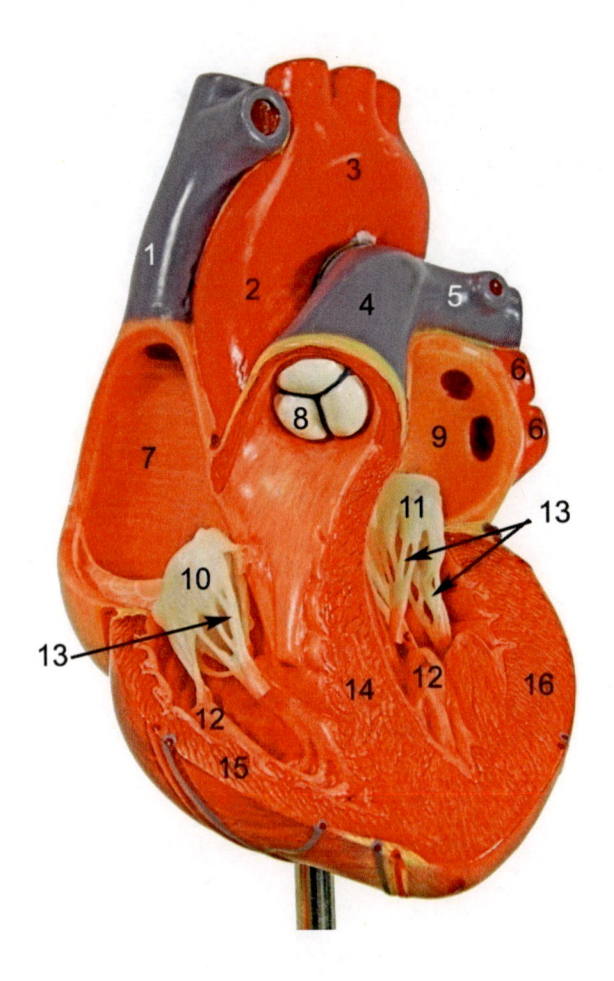

Figure 5-29. Heart – anterior interior

Figure 5-30. Heart – left side
(Copyright Somso)

1. Superior vena cava
2. Ascending aorta
3. Aortic arch
4. Pulmonary trunk
5. Left pulmonary artery
6. Left pulmonary veins
7. Right atrium
8. Pulmonary valve
9. Left atrium
10. Tricuspid valve
11. Mitral (bicuspid) valve
12. Papillary muscles
13. Chordae tendineae
14. Interventricular septum
15. Lateral wall of right ventricle
16. Lateral wall of left ventricle
17. Descending aorta
18. Aortic valve

112

Chapter 6
Respiratory System Models

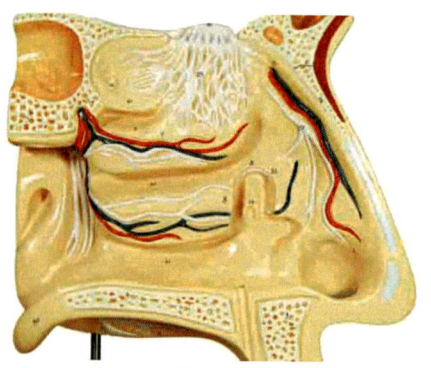

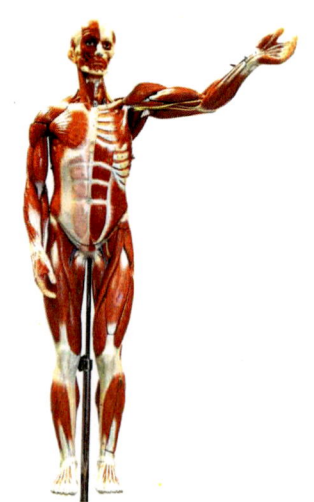

Male Muscle Figure
(Copyright Somso)

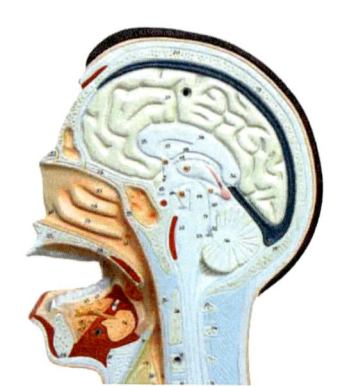

Nose
(Copyright Denoyer-Geppert)

Head and Neck
(Copyright Denoyer-Geppert)

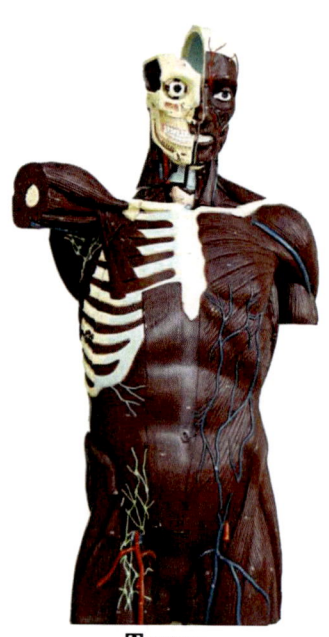

Torso
(Copyright Somso)

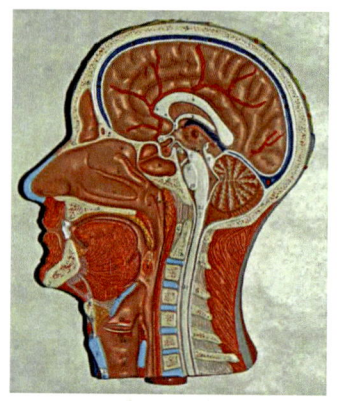

Head and Neck
(Copyright 3B Scientific)

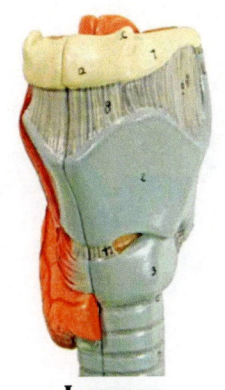

Larynx
(Copyright Somso)

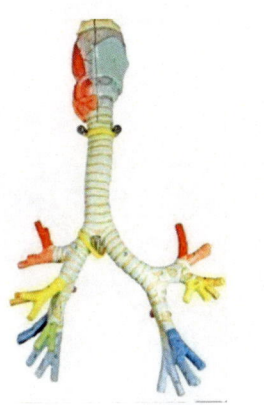

Larynx and Trachea
(Copyright Somso)

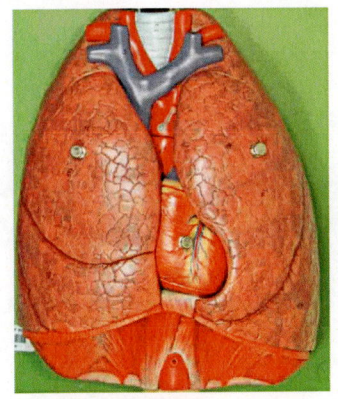

Lungs
(Copyright Somso)

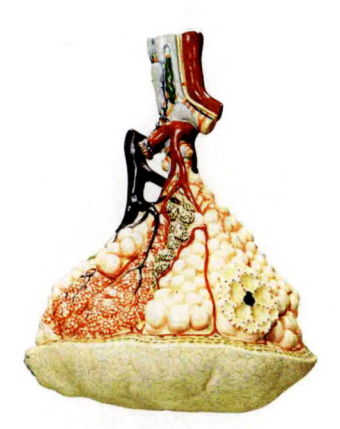

Lung Lobule
(Copyright Somso)

Male Muscle Figure

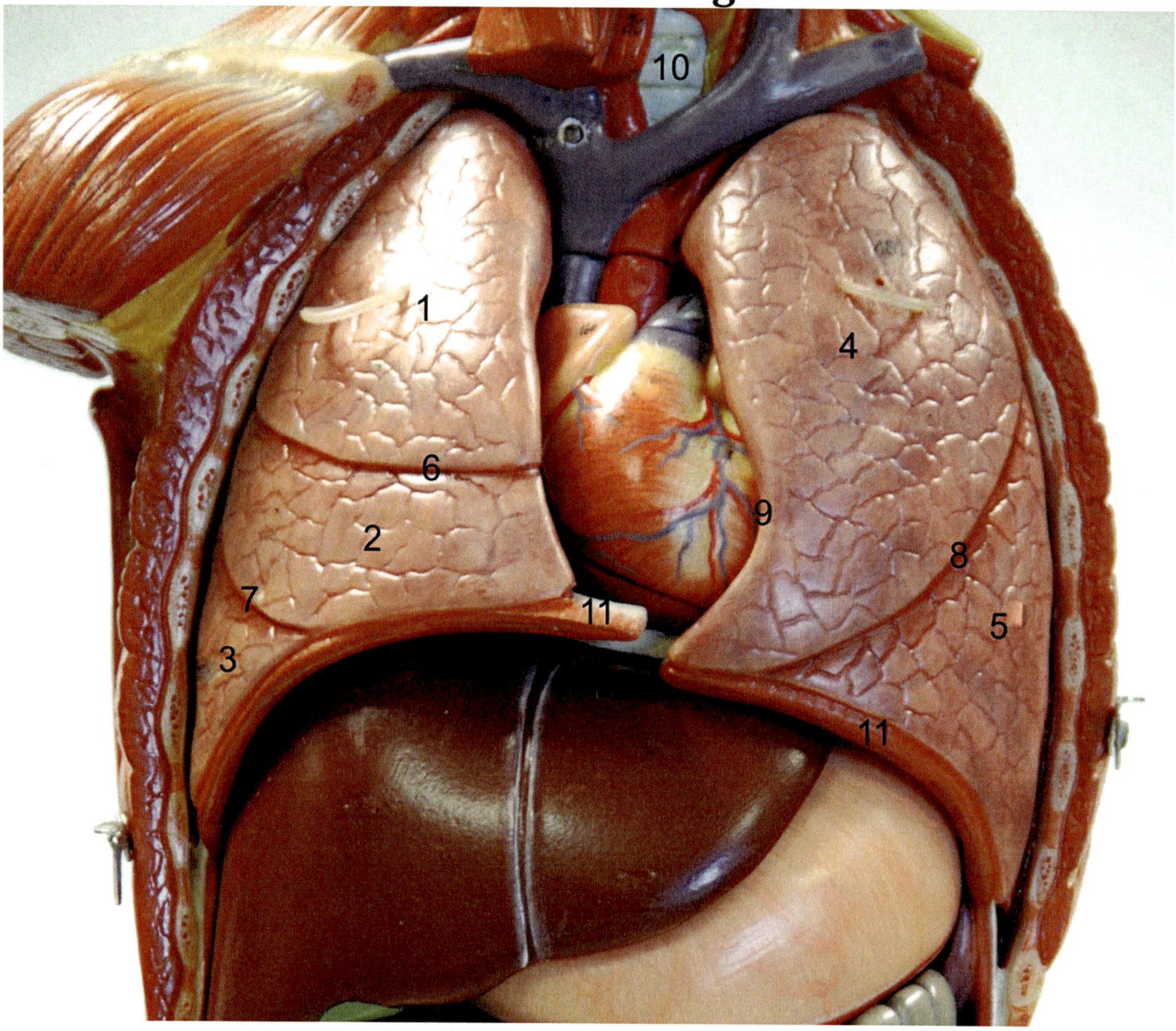

Figure 6-1. Respiratory Organs
(Copyright Somso)

1. Superior right lobe
2. Middle right lobe
3. Inferior right lobe
4. Superior left lobe
5. Inferior left lobe
6. Horizontal fissure
7. Oblique fissure
8. Oblique fissure
9. Cardiac notch
10. Trachea
11. Diaphragm

Torso – Somso

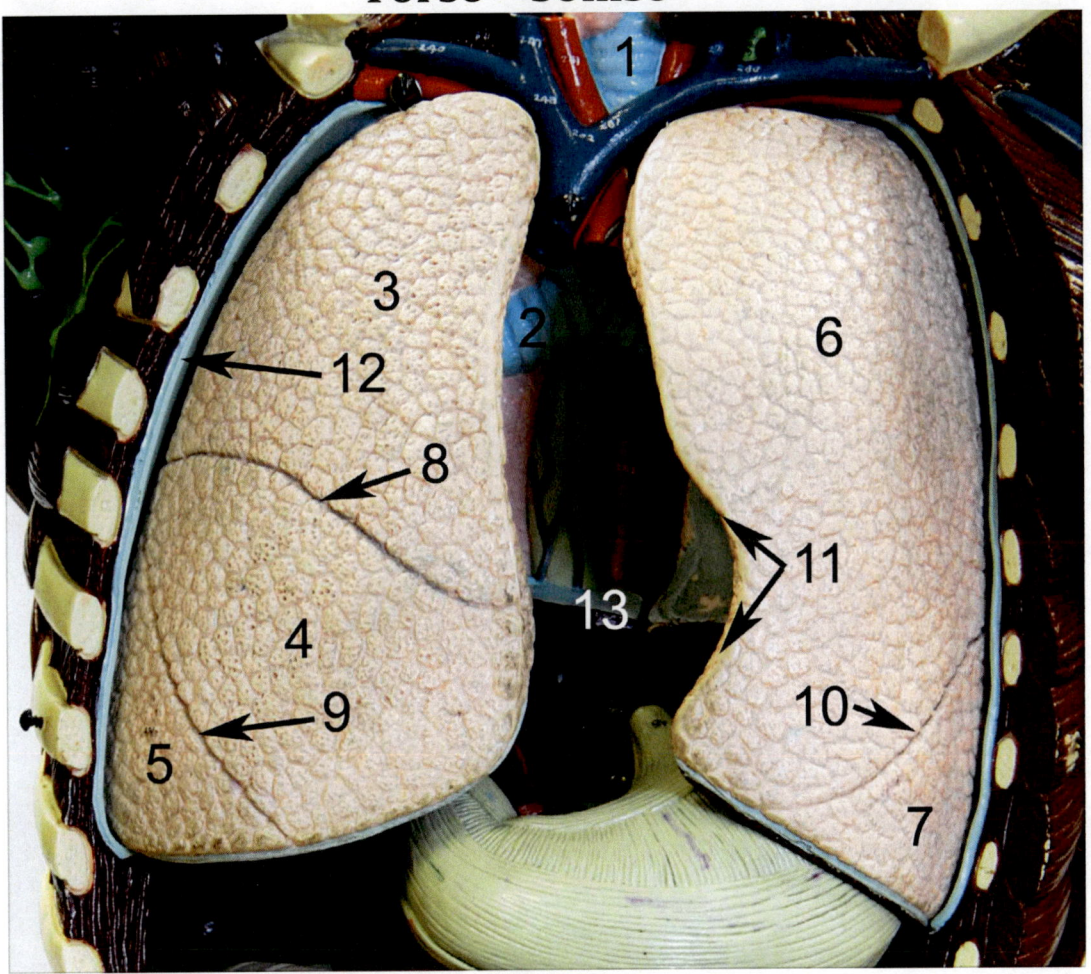

Figure 6-2. Respiratory System
(Copyright Somso)
1. Trachea
2. Primary bronchus
3. Superior right lobe
4. Middle right lobe
5. Inferior right lobe
6. Superior left lobe
7. Inferior left lobe
8. Horizontal fissure
9. Oblique fissure
10. Oblique fissure
11. Cardiac notch
12. Parietal pleura
13. Diaphragm

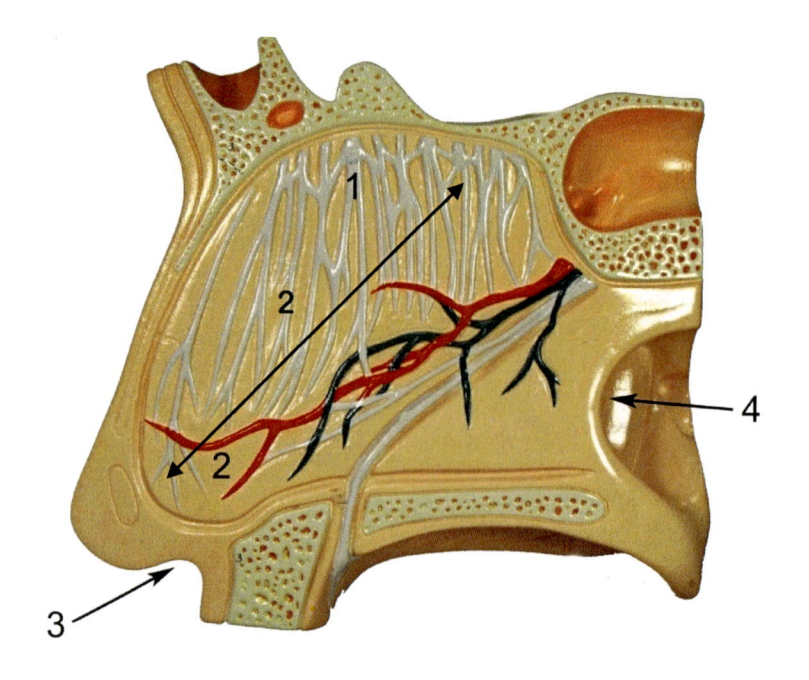

Figure 6-3. Nose - septum

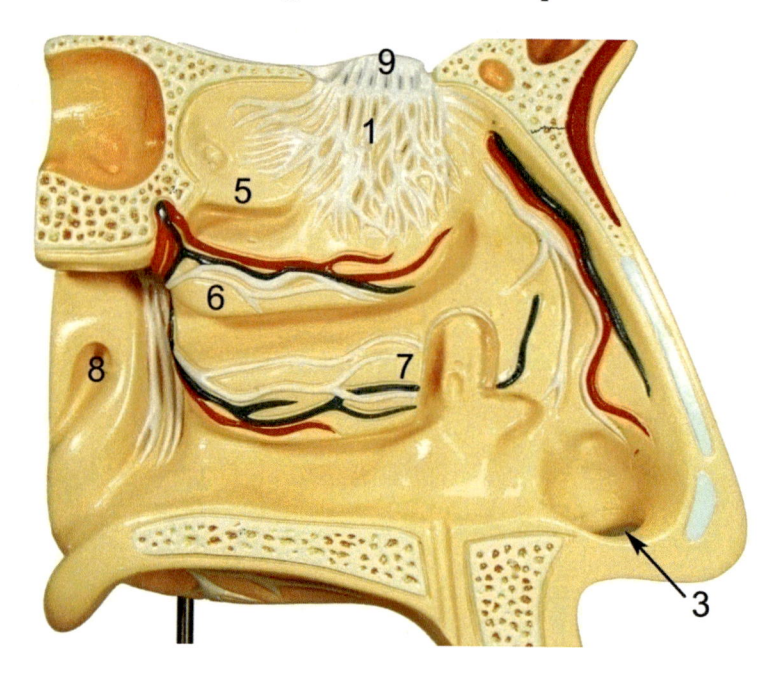

Figure 6-4. Nose - turbinates
(Copyright Denoyer-Geppert)

1. Olfactory neurons
2. Nasal septum
3. External naris
4. Internal naris
5. Superior nasal concha (turbinate)
6. Middle nasal concha (turbinate)
7. Inferior nasal concha (turbinate)
8. Opening of auditory (Eustachian) tube
9. Olfactory bulb

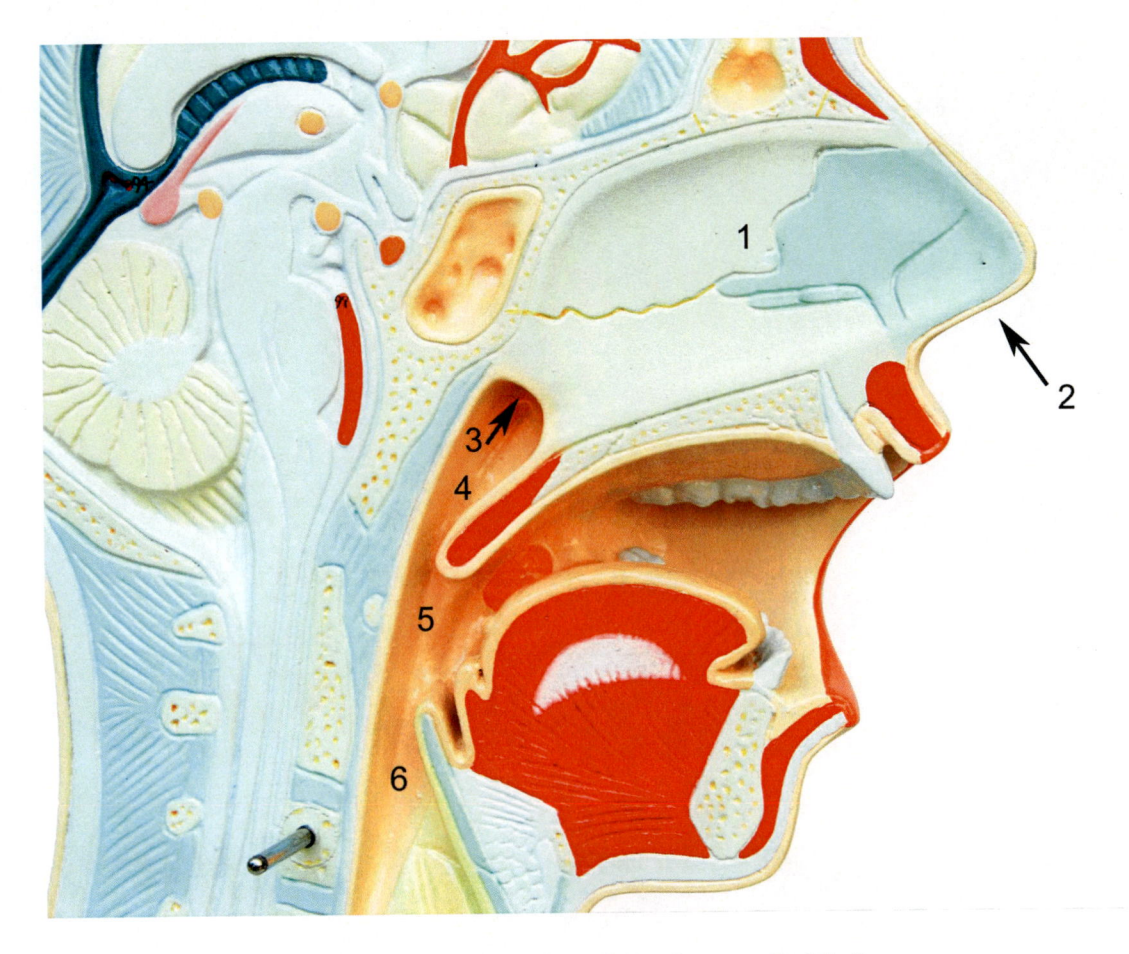

Figure 6-5. Head and Neck – medial left
(Copyright Denoyer-Geppert)

1. Nasal septum
2. External naris
3. Internal naris
4. Nasopharynx
5. Oropharynx
6. Laryngopharynx

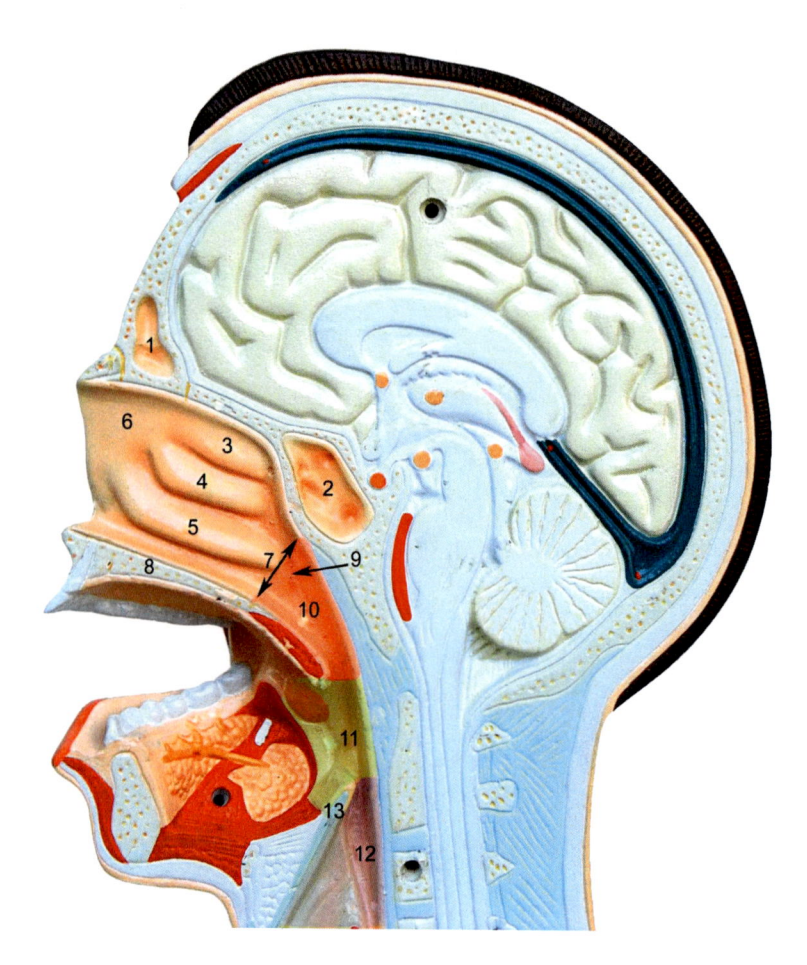

Figure 6-6. Head and Neck – medial right
(Copyright Denoyer-Geppert)

1. Frontal sinus
2. Sphenoid sinus
3. Superior nasal concha (turbinate)
4. Middle nasal concha (turbinate)
5. Inferior nasal concha (turbinate)
6. Nasal cavity
7. Internal naris
8. Hard palate
9. Opening to auditory (Eustachian) tube
10. Nasopharynx (orange)
11. Oropharynx (green)
12. Laryngopharynx (purple)
13. Epiglottis

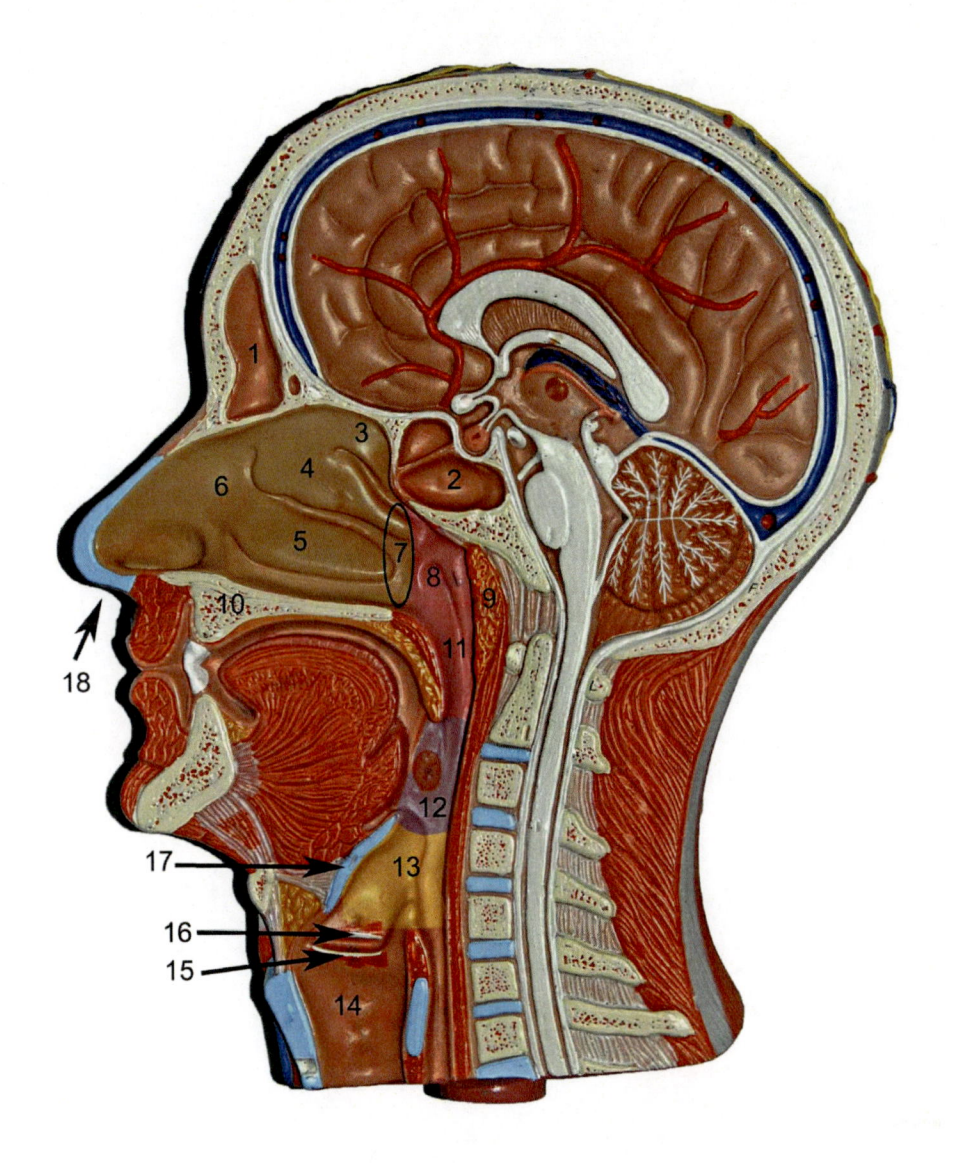

Figure 6-7. Head and Neck – medial
(Copyright 3B Sceintific)

1. Frontal sinus
2. Sphenoid sinus
3. Superior nasal concha (turbinate)
4. Middle nasal concha (turbinate)
5. Inferior nasal concha (turbinate)
6. Nasal cavity
7. Internal naris
8. Opening to auditory (Eustachian) tube
9. Pharyngeal tonsil
10. Hard palate
11. Nasopharynx (pink)
12. Oropharynx (purple)
13. Laryngopharynx (yellow)
14. Larynx
15. Vocal fold or cord
16. Vestibular fold
17. Epiglottis
18. External naris

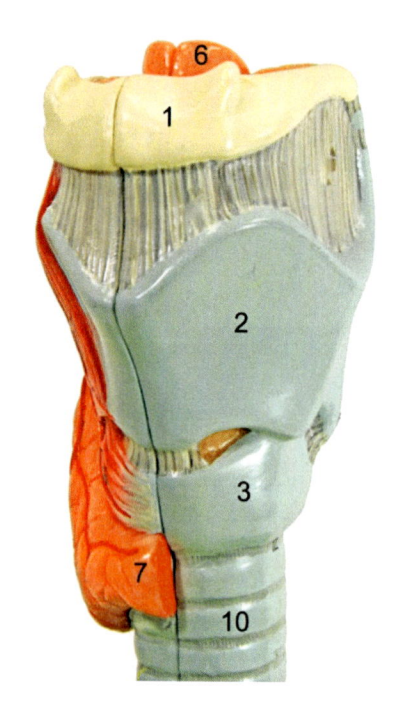

Figure 6-8. Larynx - anterior

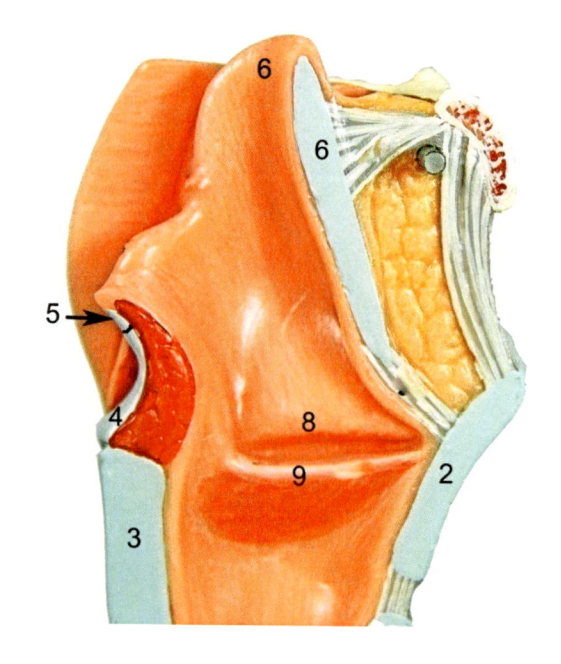

Figure 6-10. Larynx - sagittal

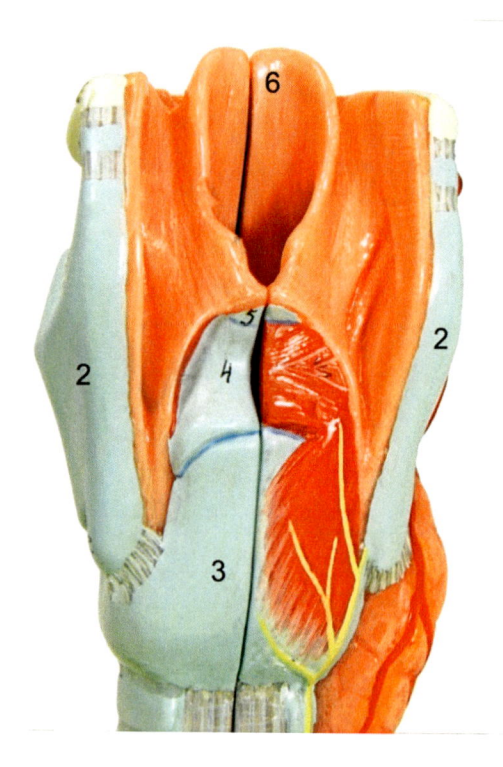

Figure 6-9. Larynx - posterior
(Copyright Somso)

1. Hyoid bone
2. Thyroid cartilage
3. Cricoid cartilage
4. Arytenoid cartilage
5. Corniculate cartilage
6. Epiglottis
7. Thyroid gland
8. Vestibular fold
9. Vocal fold or cord
10. Tracheal cartilage

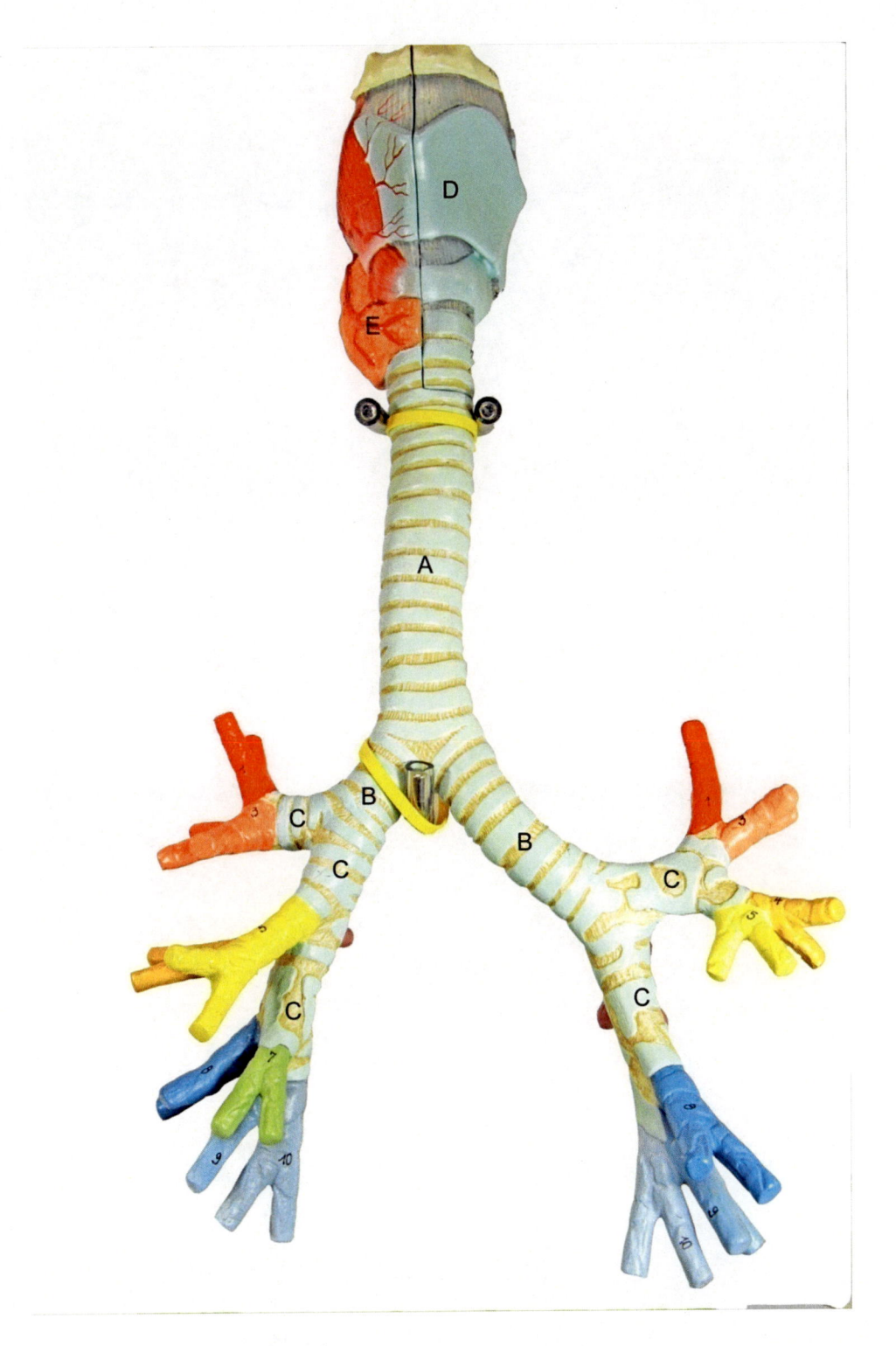

Figure 6-11. Larynx and Trachea
(Copyright Somso)

A. Trachea
B. Primary bronchus
C. Secondary bronchus
D. Larynx
E. Thyroid gland
1-10 Tertiary (segmental) bronchi

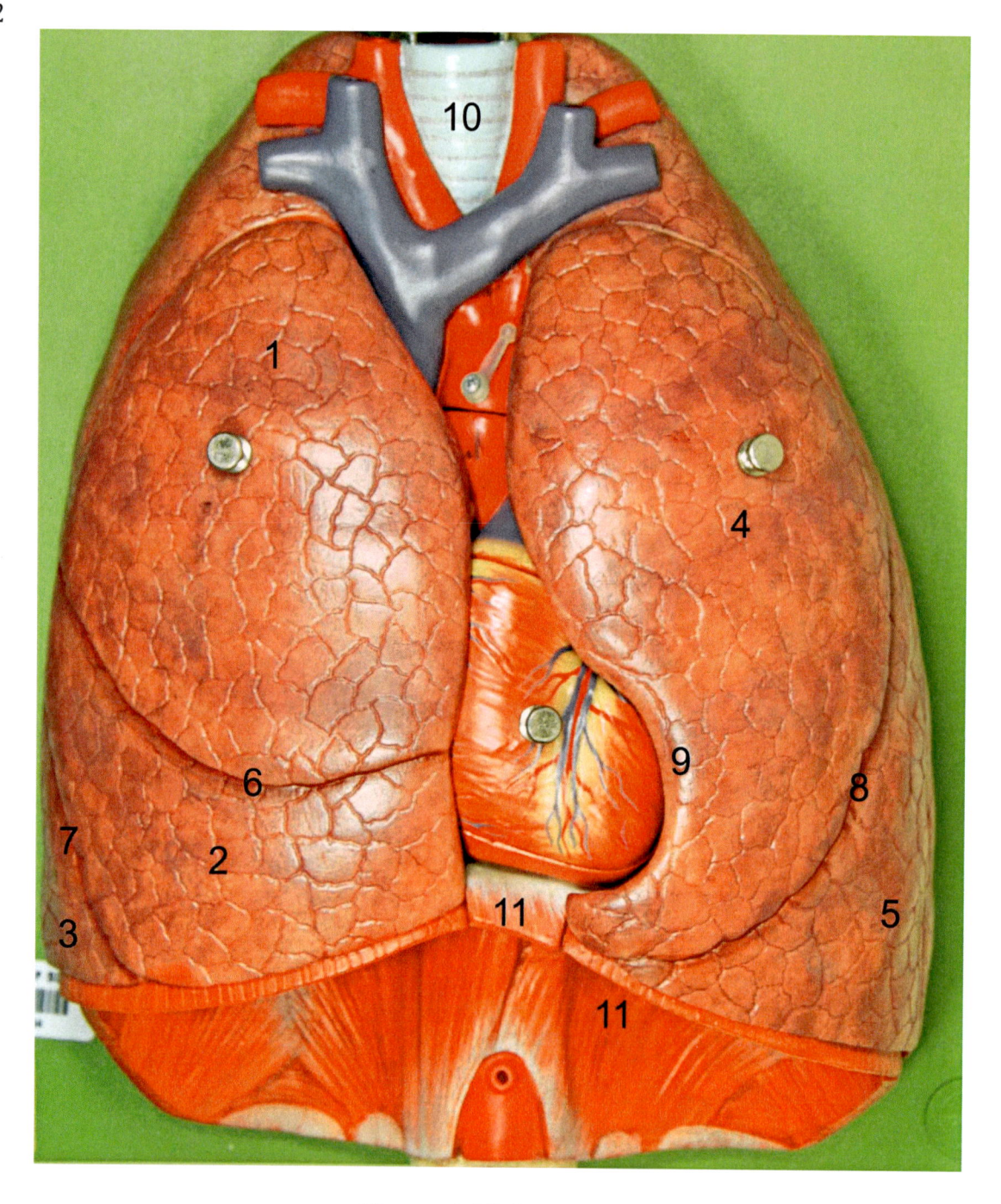

Figure 6-12. Lungs
(Copyright Somso)

1. Superior right lobe
2. Middle right lobe
3. Inferior right lobe
4. Superior left lobe
5. Inferior left lobe
6. Horizontal fissure
7. Oblique fissure
8. Oblique fissure
9. Cardiac notch
10. Trachea
11. Diaphragm

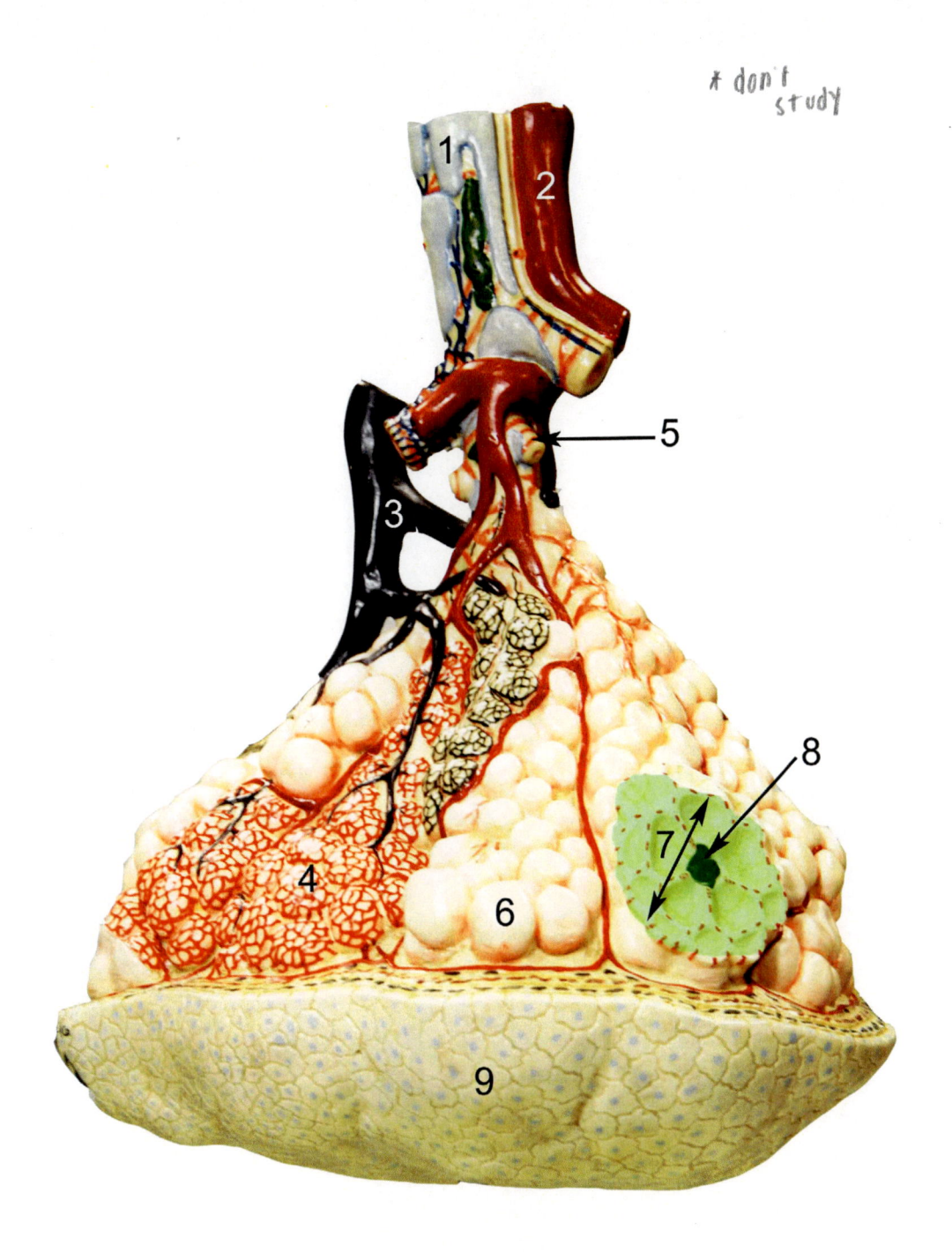

Figure 6-13. Lung Lobule
(Copyright Somso)

1. Small bronchus
2. Pulmonary venule
3. Pulmonary arteriole
4. Pulmonary capillaries
5. Bronchiole
6. Alveolus
7. Alveolar sac
8. Alveolar duct (end)
9. Visceral pleura

Chapter 7
Digestive System Models

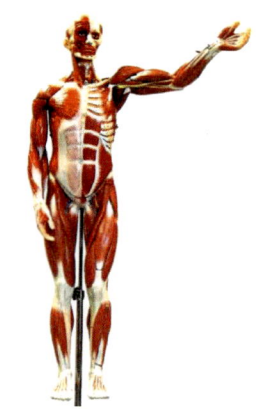

Male Muscle Figure
(Copyright Somso)

Stomach
(Copyright Somso)

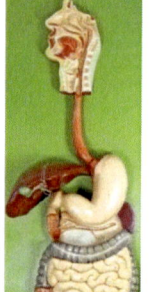

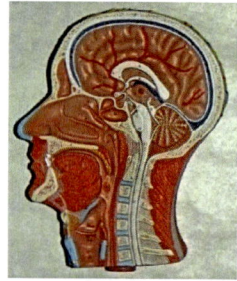

Head and Neck
(Copyright 3B Scientific)

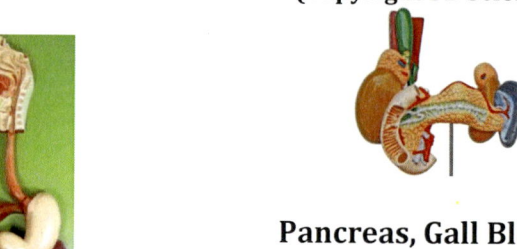

Pancreas, Gall Bladder and Duodenum
(Copyright Denoyer-Geppert)

Torso
(Copyright 3B Scientific)

Digestive System
(Copyright Somso)

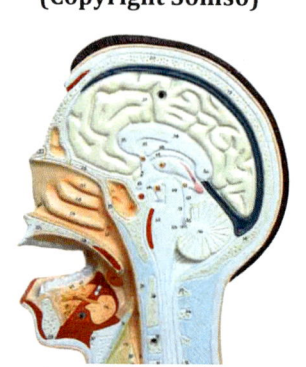

Teeth
(Copyright Somso)

Torso
(Copyright Somso)

Head and Neck
(Copyright Denoyer-Geppert)

Intestinal Villi
(Copyright Denoyer-Geppert)

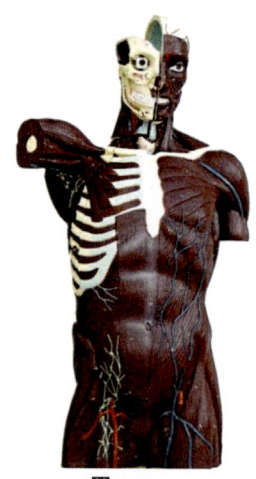

Male Muscle Figure

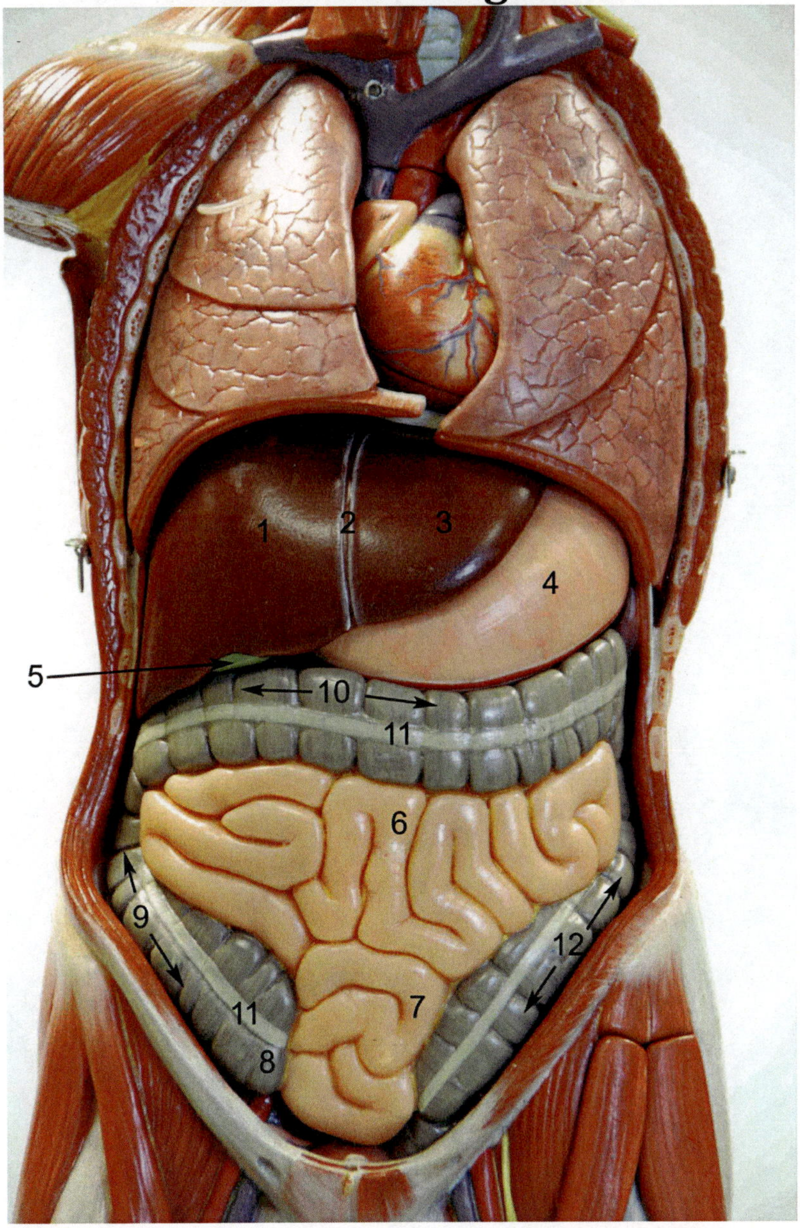

Figure 7-1. Digestive Organs
(Copyright Somso)

1. Right lobe of liver
2. Falciform ligament
3. Left lobe of liver
4. Stomach
5. Gall bladder
6. Jejunum
7. Ileum
8. Cecum
9. Ascending colon
10. Transverse colon
11. Taenia coli
12. Descending colon

Torso – 3B

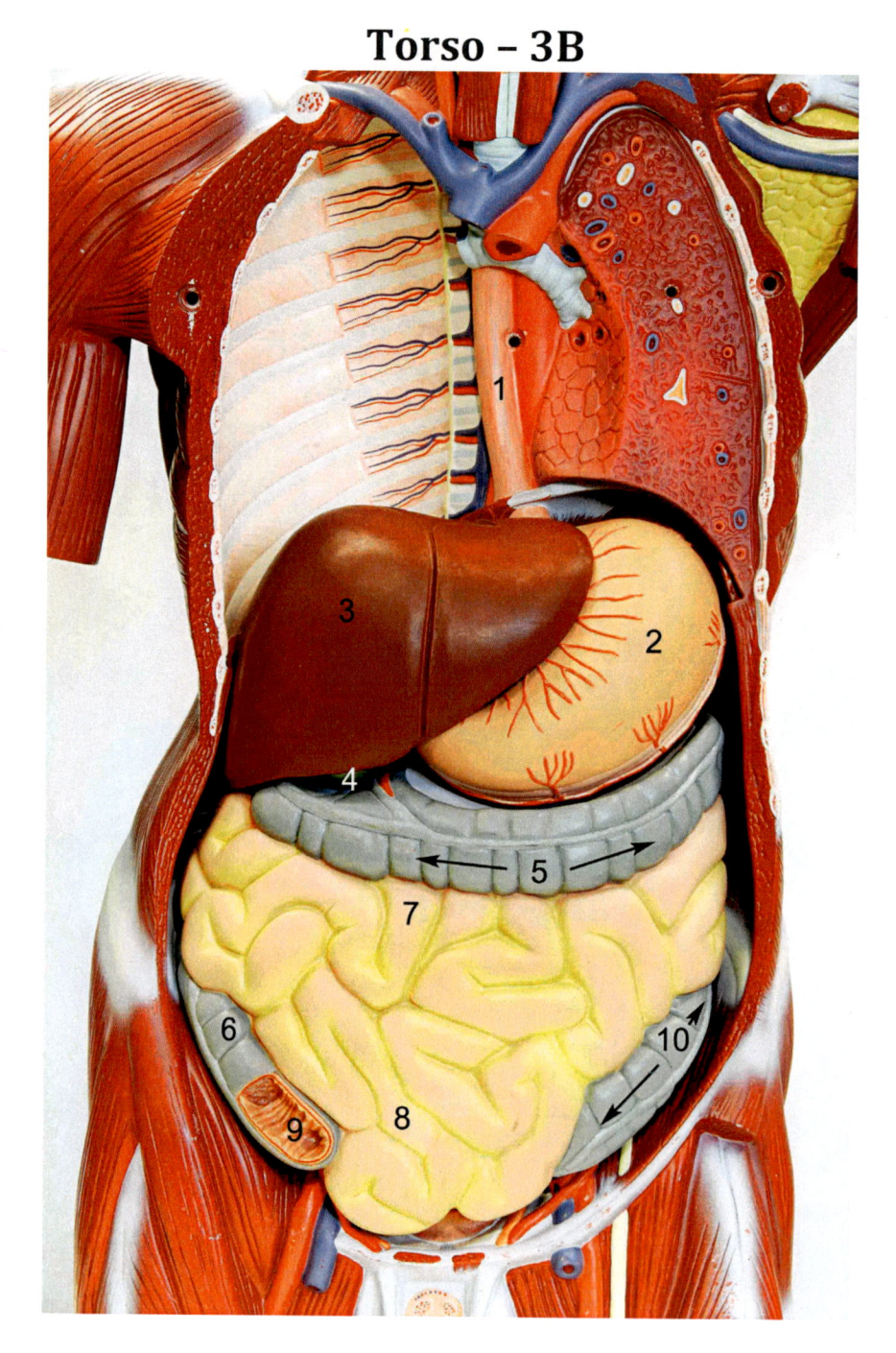

Figure 7-2. Digestive Organs
(Copyright 3B Scientific)

1. Esophagus
2. Stomach
3. Liver
4. Gall bladder
5. Transverse colon
6. Ascending colon
7. Jejunum
8. Ileum
9. Cecum
10. Descending colon

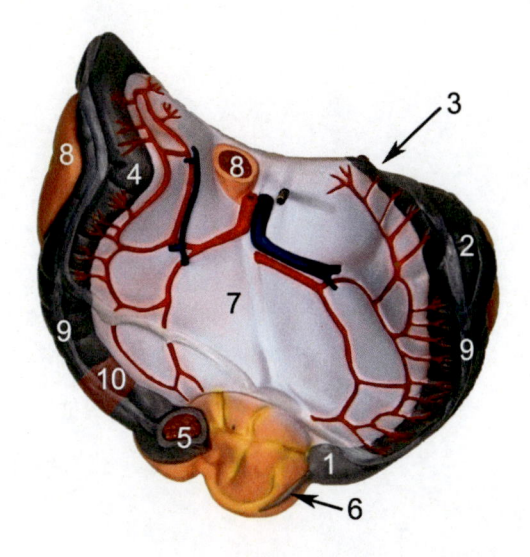

Figure 7-3. Torso – Intestines - posterior view

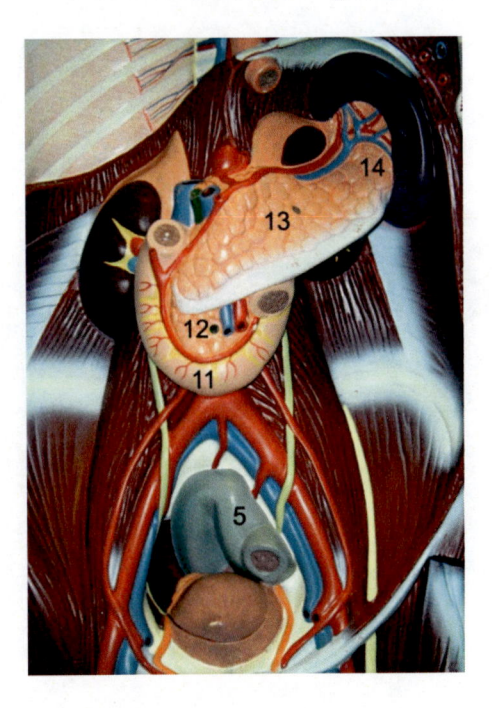

Figure 7-4. Torso – Digestive Organs - deep
(Copyright 3B Scientific)

1. Cecum
2. Ascending colon
3. Transverse colon
4. Descending colon
5. Sigmoid colon
6. Appendix
7. Mesocolon
8. Small intestine
9. Taenia coli
10. Haustrum
11. Duodenum
12. Head of pancreas
13. Body of pancreas
14. Tail of pancreas

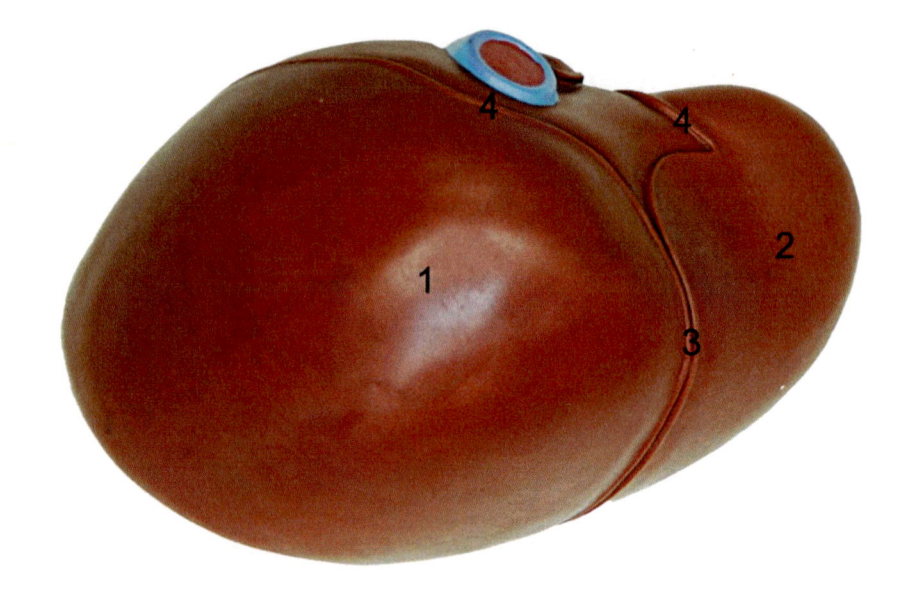

Figure 7-5. Liver – anterior/superior

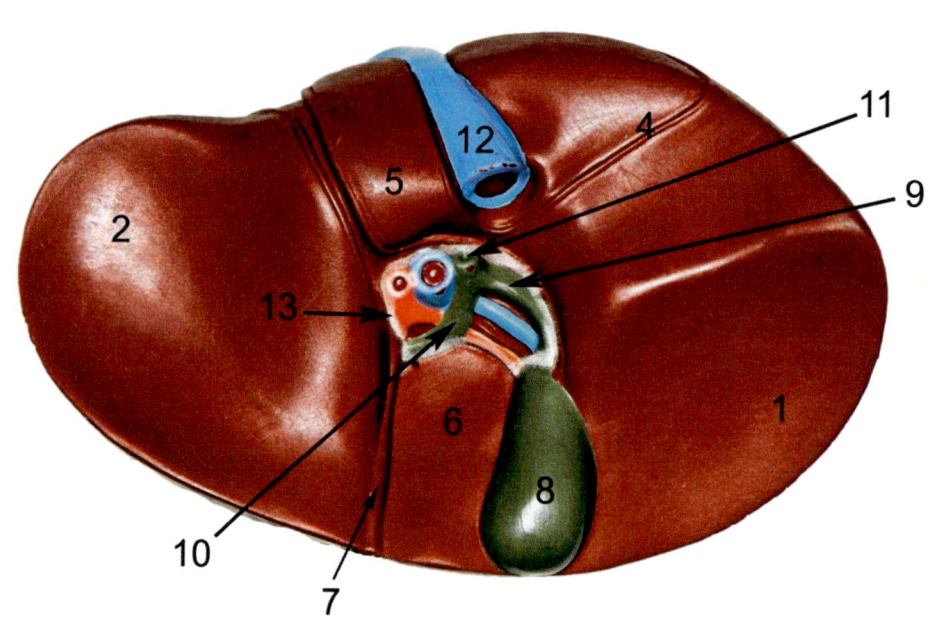

Figure 7-6. Liver – inferior
(Copyright 3b Scientific)

1. Right lobe
2. Left lobe
3. Falciform ligament
4. Coronary ligament
5. Caudate lobe
6. Quadrate lobe
7. Round ligament
8. Gall bladder
9. Cystic duct
10. Hepatic duct
11. Common bile duct
12. Inferior vena cava
13. Porta hepatis or Hilum (ringed in white)

don't study

Torso - Somso

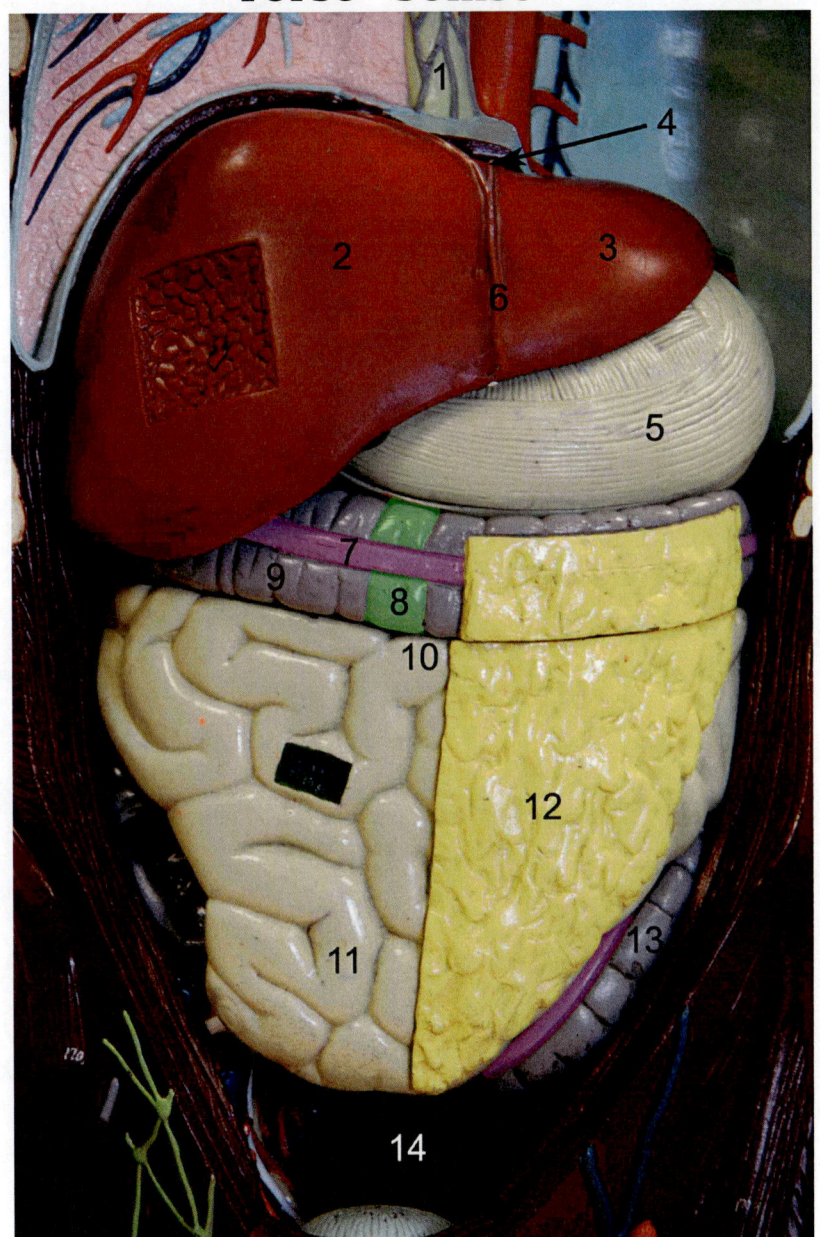

Figure 7-7. Digestive Organs
(Copyright Somso)

1. Esophagus
2. Right lobe of liver
3. Left lobe of liver
4. Coronary ligament
5. Stomach
6. Falciform ligament
7. Taenia coli
8. Haustrum
9. Transverse colon
10. Jejunum
11. Ileum
12. Greater omentum
13. Descending colon
14. Rectum

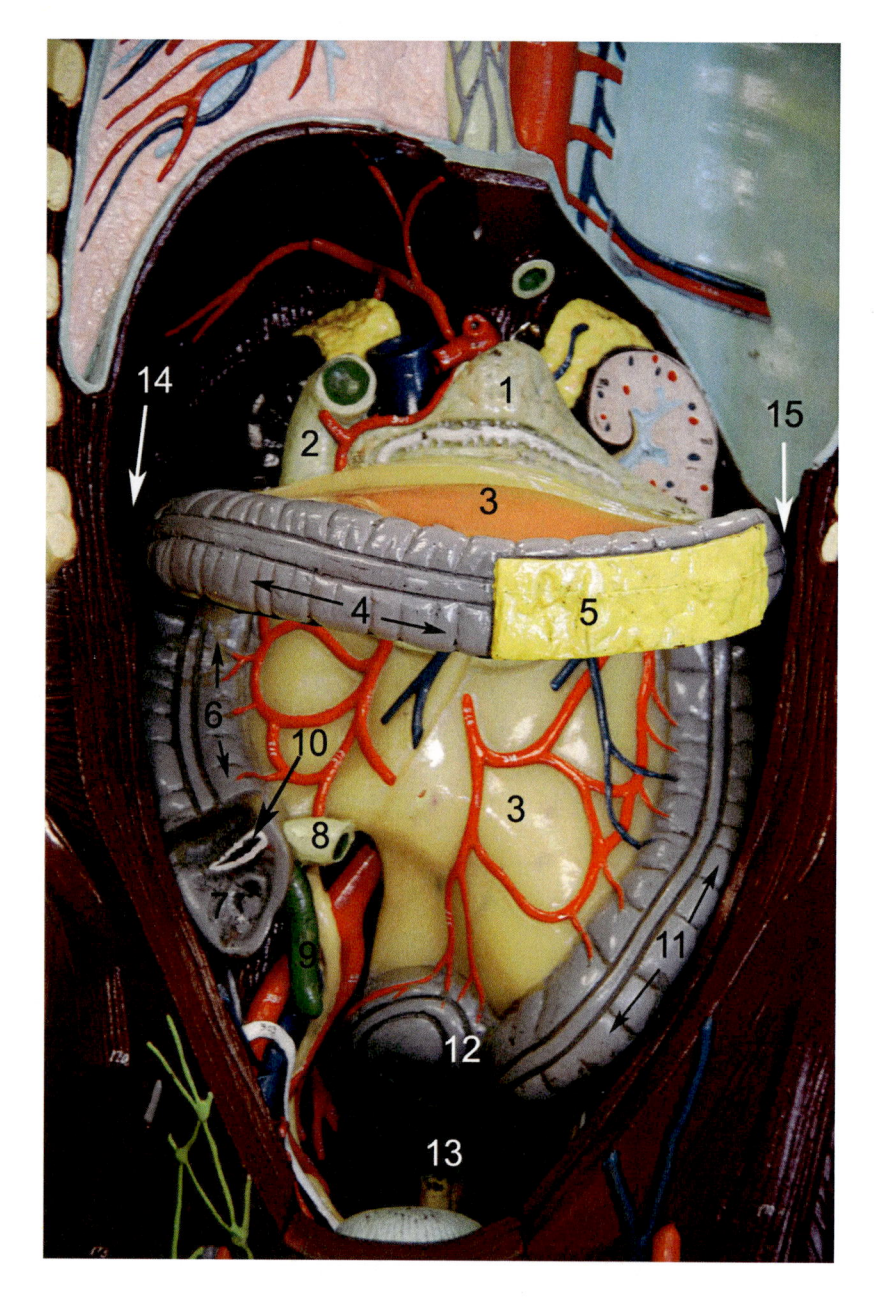

Figure 7-8. Digestive Organs – deep
(Copyright Somso)

1. Pancreas
2. Duodenum
3. Mesocolon
4. Transverse colon
5. Greater omentum
6. Ascending colon
7. Cecum
8. Ileum
9. Appendix
10. Ileocecal valve
11. Descending colon
12. Sigmoid colon
13. Rectum
14. Right colic (hepatic) flexure
15. Left colic (splenic) flexure

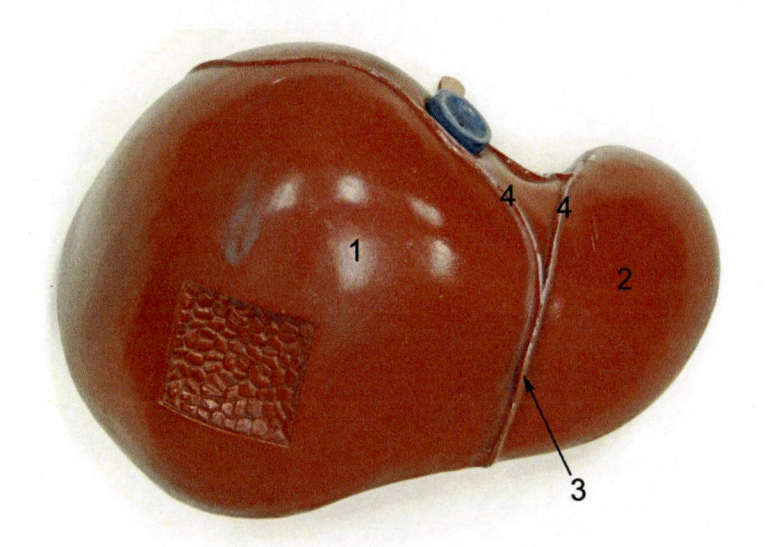

Figure 7-9. Liver - anterior

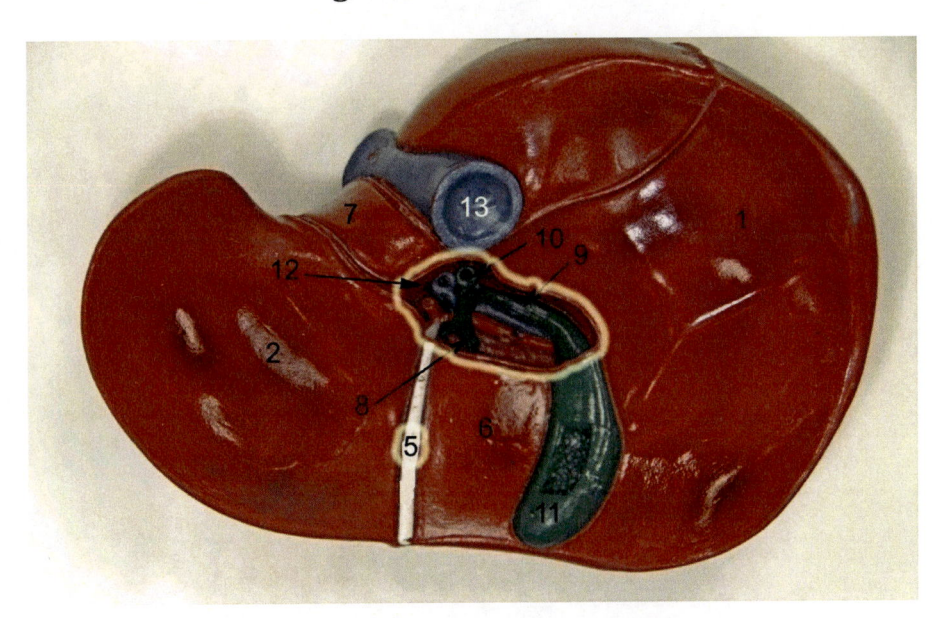

Figure 7-10. Liver - inferior
(Copyright Somso)

1. Right lobe
2. Left lobe
3. Falciform ligament
4. Coronary ligament
5. Round ligament
6. Quadrate lobe
7. Caudate lobe
8. Hepatic duct
9. Cystic duct
10. Common bile duct
11. Gall bladder
12. Porta hepatis or Hilum (ringed in white)
13. Inferior vena cava

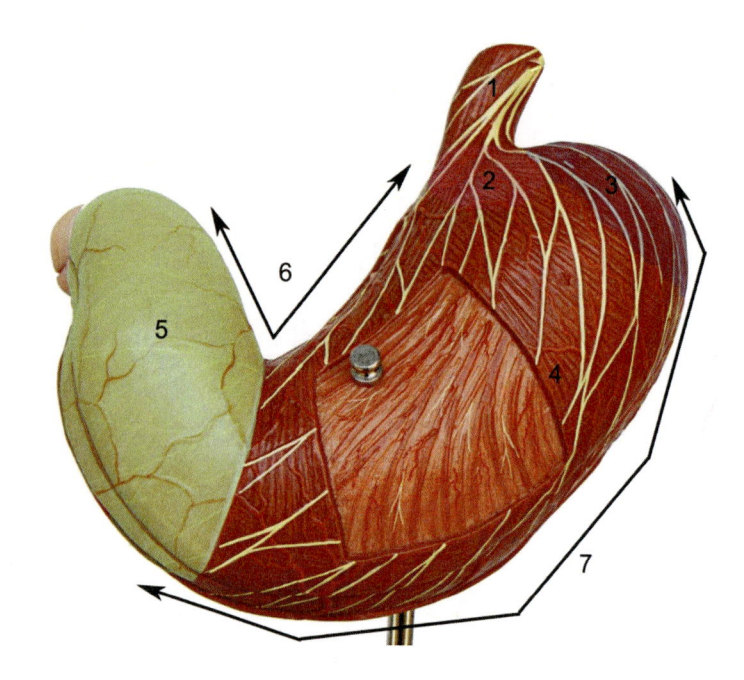

Figure 7-11. Stomach

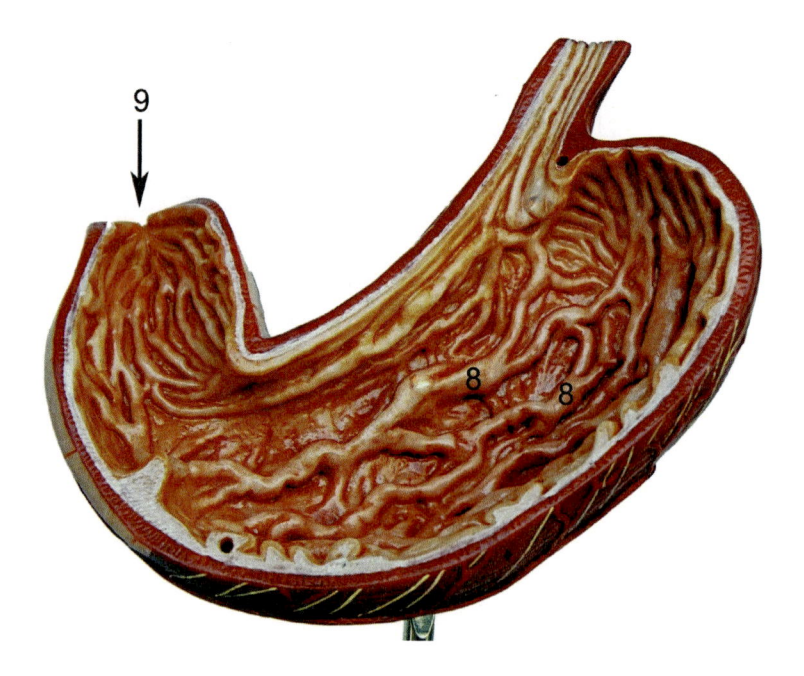

Figure 7-12. Stomach - interior
(Copyright Somso)

1. Esophagus
2. Cardia of stomach
3. Fundus of stomach
4. Body of stomach
5. Pylorus of stomach
6. Lesser curvature
7. Greater curvature
8. Rugae
9. Pyloric sphincter

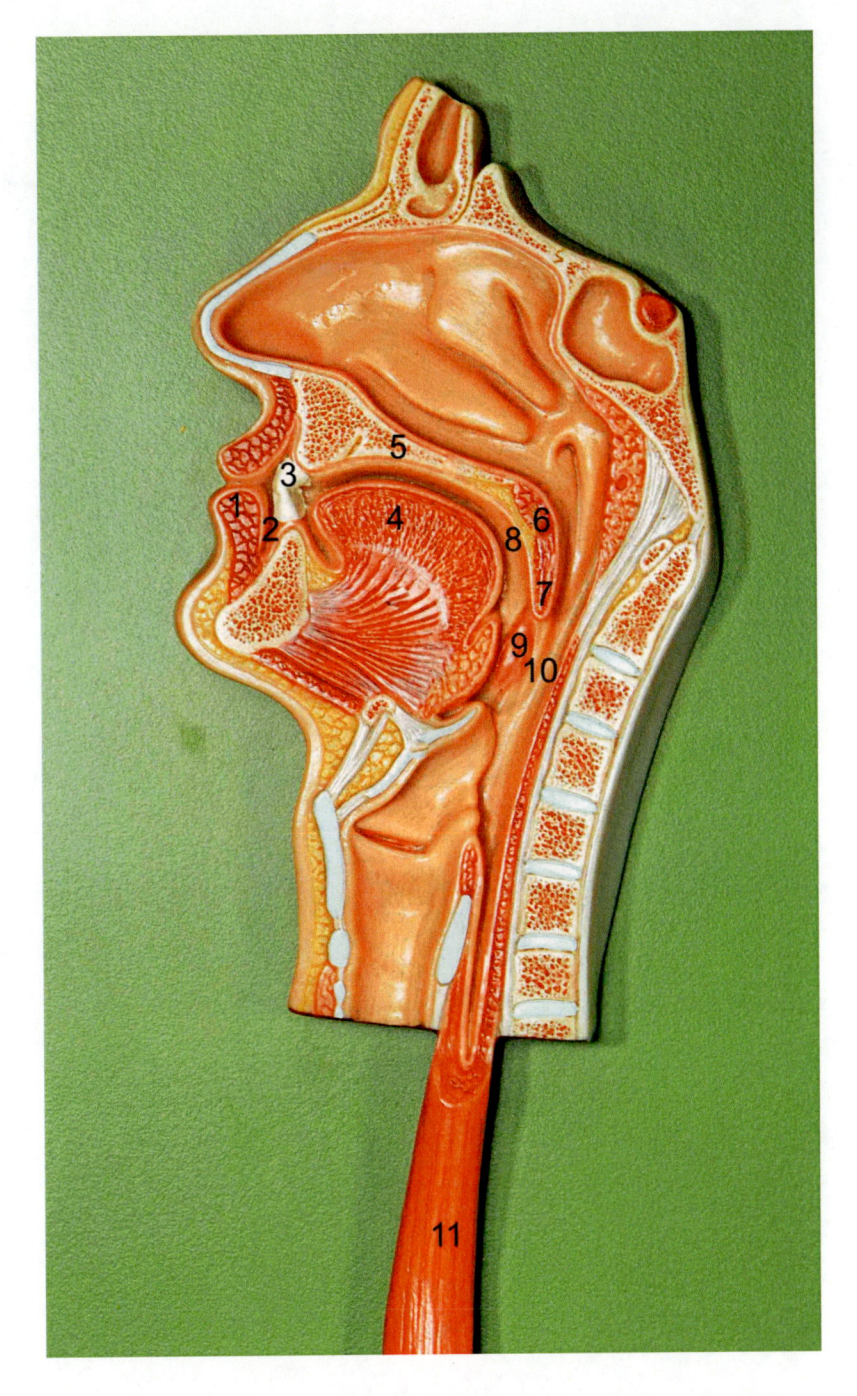

Figure 7-13. Digestive System - head and neck
(Copyright Somso)

1. Lips
2. Vestibule
3. Teeth
4. Tongue
5. Hard palate
6. Soft palate
7. Uvula
8. Oral cavity
9. Palatine tonsil
10. Oropharynx
11. Esophagus

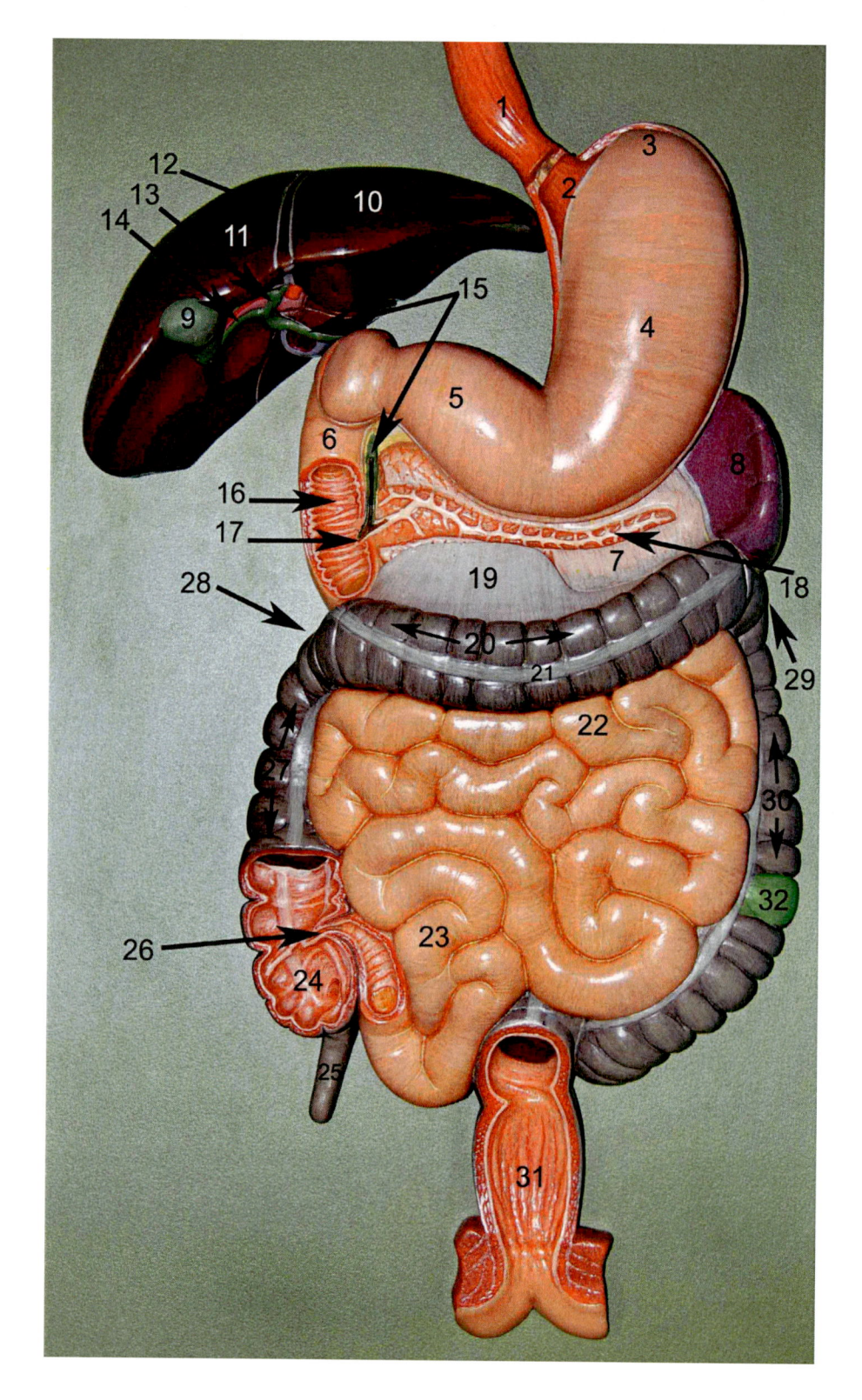

Figure 7-14. Digestive System
(Copyright Somso)

Identification Key for Figure 7-14

1. Esophagus
2. Cardia of stomach
3. Fundus of stomach
4. Body of stomach
5. Pylorus of stomach
6. Duodenum
7. Pancreas
8. Spleen
9. Gall bladder
10. Left lobe of liver
11. Right lobe of liver
12. Falciform ligament
13. Hepatic duct
14. Cystic duct
15. Common bile duct
16. Plica circulares (circular fold)
17. Major duodenal papilla
18. Pancreatic duct
19. Mesocolon
20. Transverse colon
21. Taenia coli
22. Jejunum
23. Ileum
24. Cecum
25. Appendix
26. Ileocecal valve
27. Ascending colon
28. Right colic (hepatic) flexure
29. Left colic (splenic) flexure
30. Descending colon
31. Rectum
32. Haustrum

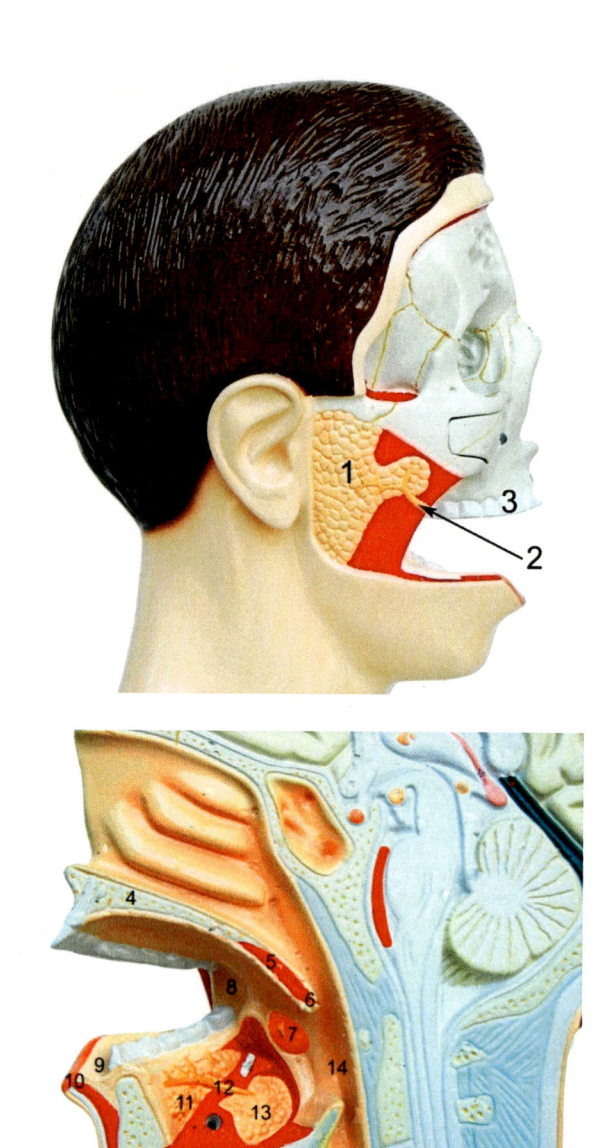

Figure 7-15. Head and Neck - lateral and medial
(Copyright Denoyer-Geppert)

1. Parotid gland
2. Parotid duct
3. Teeth
4. Hard palate
5. Soft palate
6. Uvula
7. Palatine tonsil
8. Oral cavity
9. Vestibule
10. Lips
11. Sublingual gland
12. Submandibular duct
13. Submandibular gland
14. Oropharynx
15. Esophagus

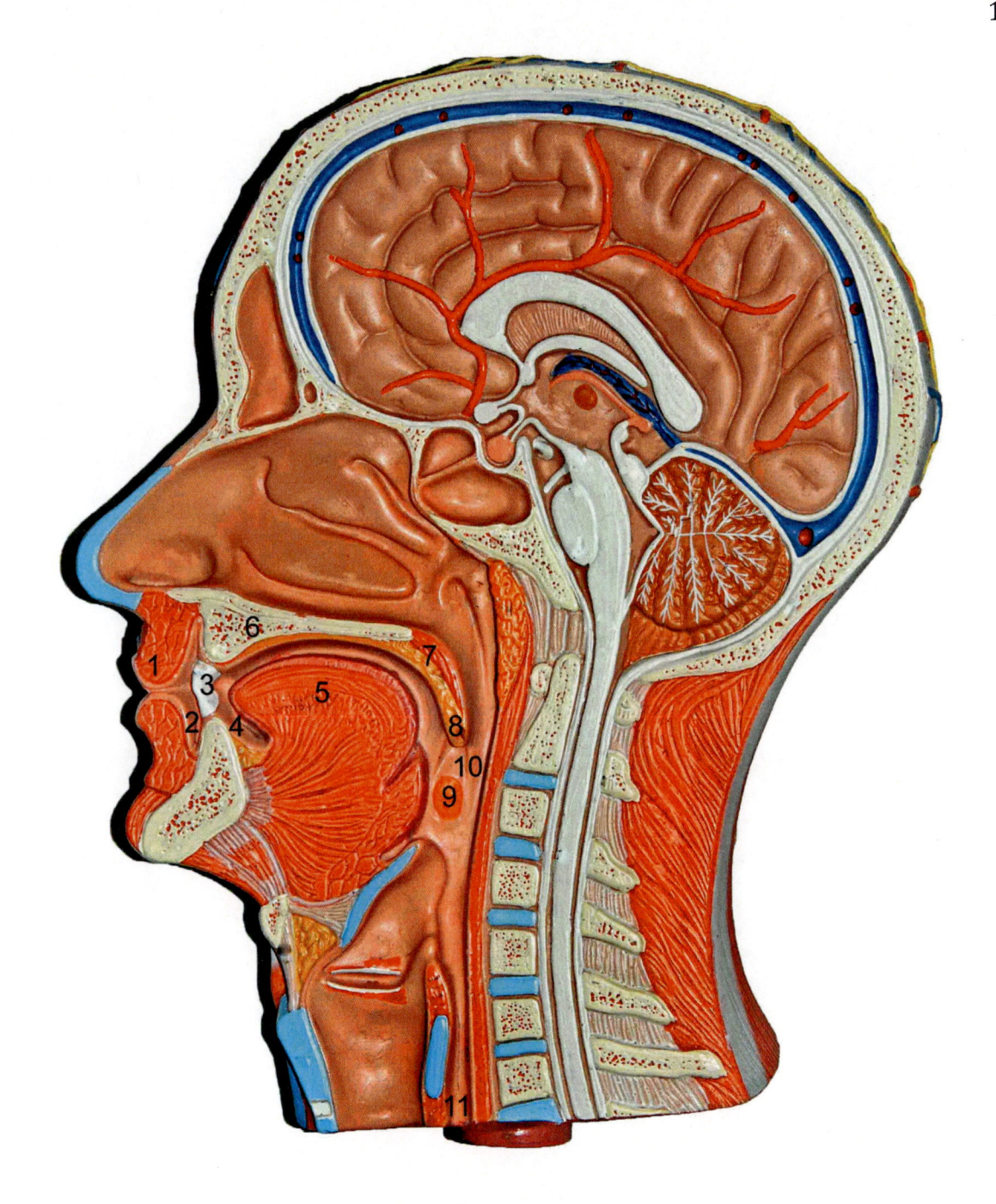

Figure 7-16. Head and Neck
(Copyright 3B Scientific)

1. Lips
2. Vestibule
3. Teeth
4. Oral cavity
5. Tongue
6. Hard palate
7. Soft palate
8. Uvula
9. Palatine tonsil
10. Oropharynx
11. Esophagus

138

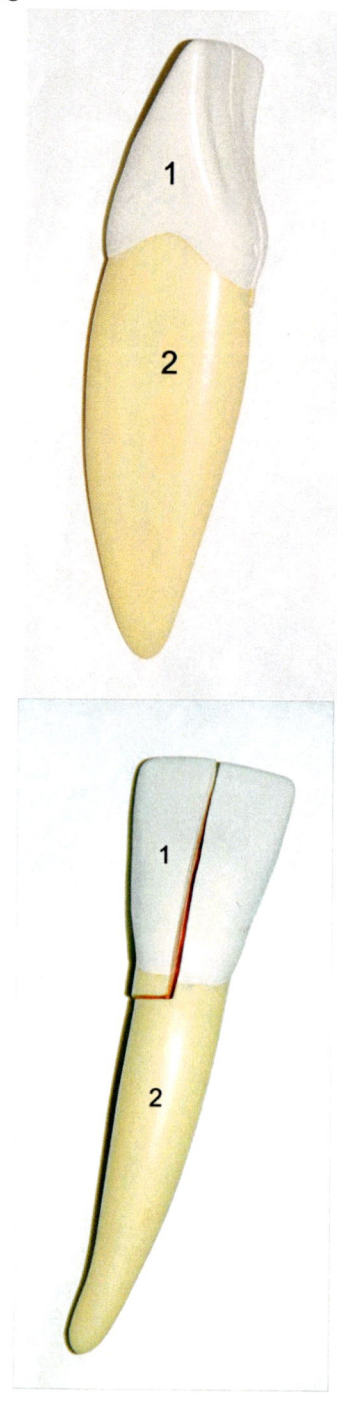

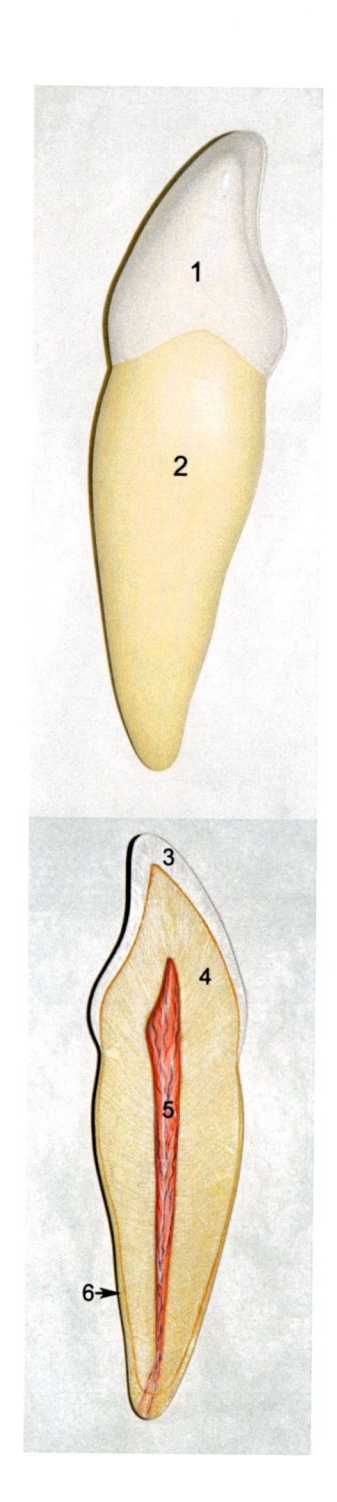

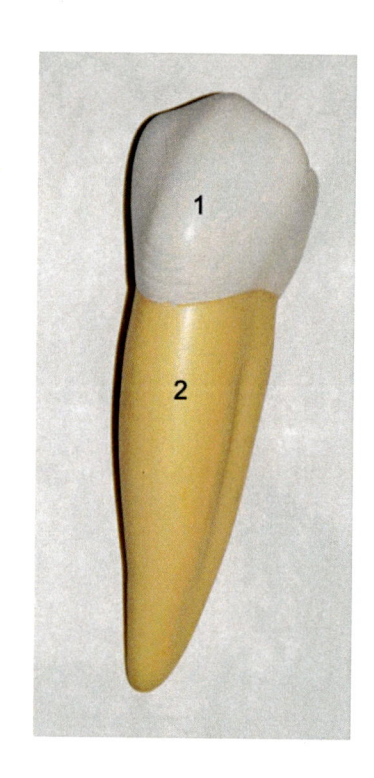

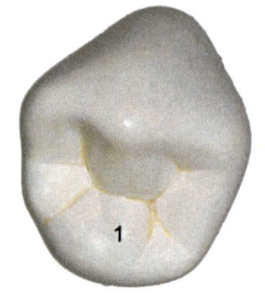

Figure 7-17. Incisor Teeth
(Copyright Somso)
1. Crown
2. Root
3. Enamel

Figure 7-18. Canine (cuspid) Teeth
4. Dentin
5. Pulp
6. Cementum

Figure 7-19. Bicuspid (premolar) Teeth

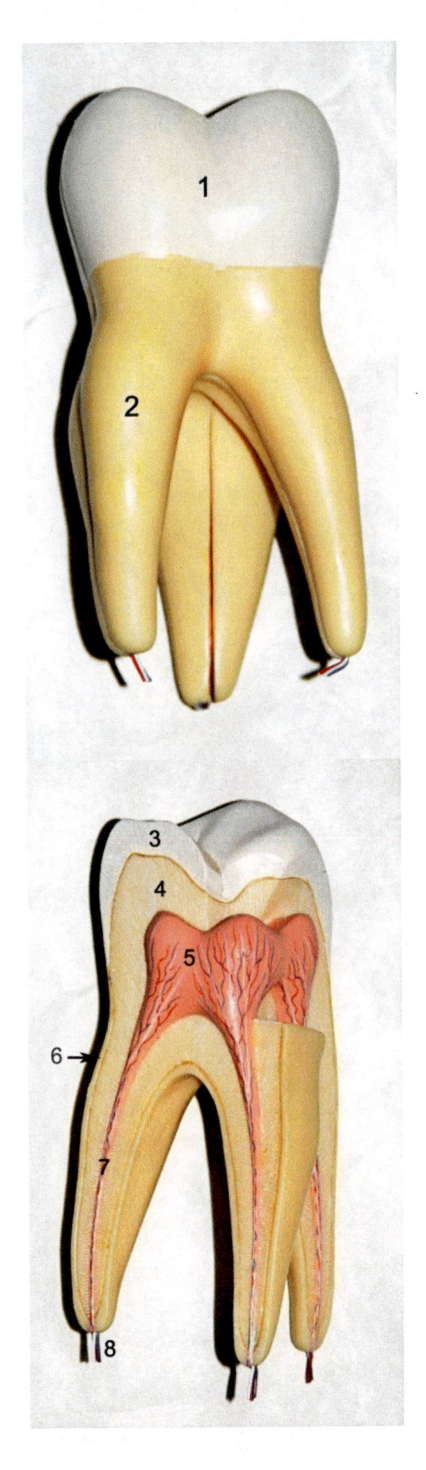

Figure 7-20. Molar Teeth
(Copyright Somso)

1. Crown
2. Root
3. Enamel
4. Dentin
5. Pulp
6. Cementum
7. Root canal
8. Nerves and vessels
9. Dental caries

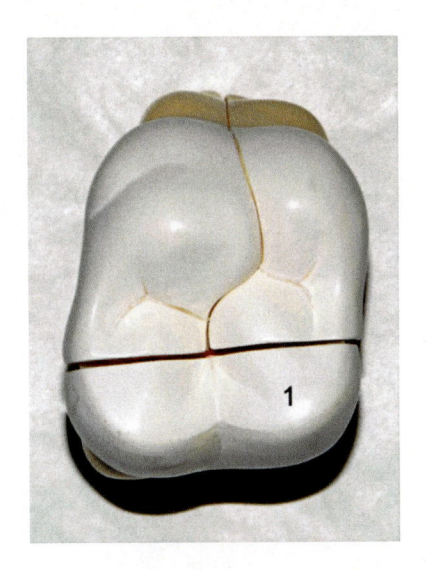

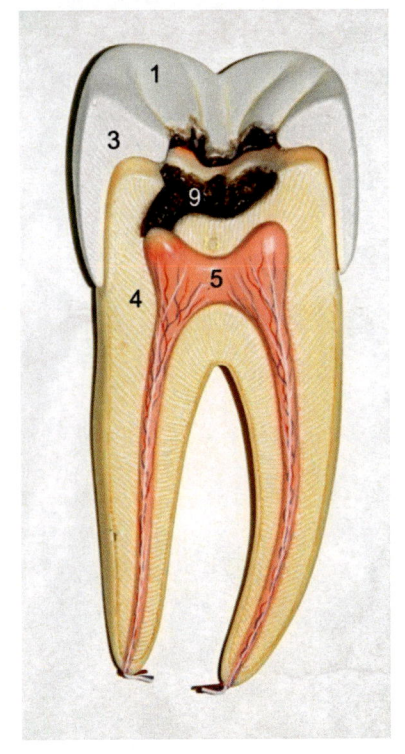

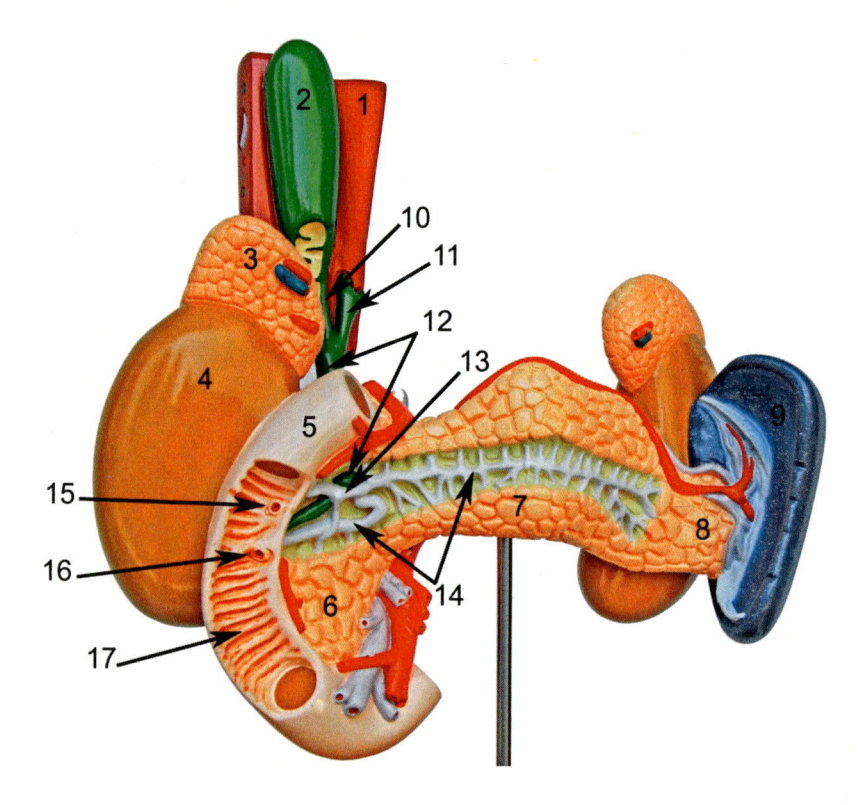

Figure 7-21. Pancreas, Gall Bladder and Duodenum

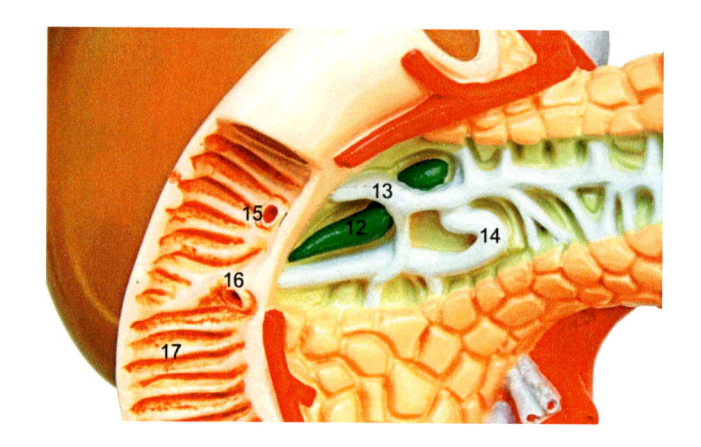

Figure 7-22. Pancreas and Duodenum – close-up
(Copyright Denoyer-Geppert)

1. Liver (section)
2. Gall bladder
3. Adrenal (suprarenal) gland
4. Kidney
5. Duodenum
6. Head of pancreas
7. Body of pancreas
8. Tail of pancreas
9. Spleen
10. Cystic duct
11. Common hepatic duct
12. Common bile duct
13. Accessory pancreatic duct
14. Pancreatic duct
15. Minor duodenal papilla
16. Major duodenal papilla
17. Plica circulares (circular fold)

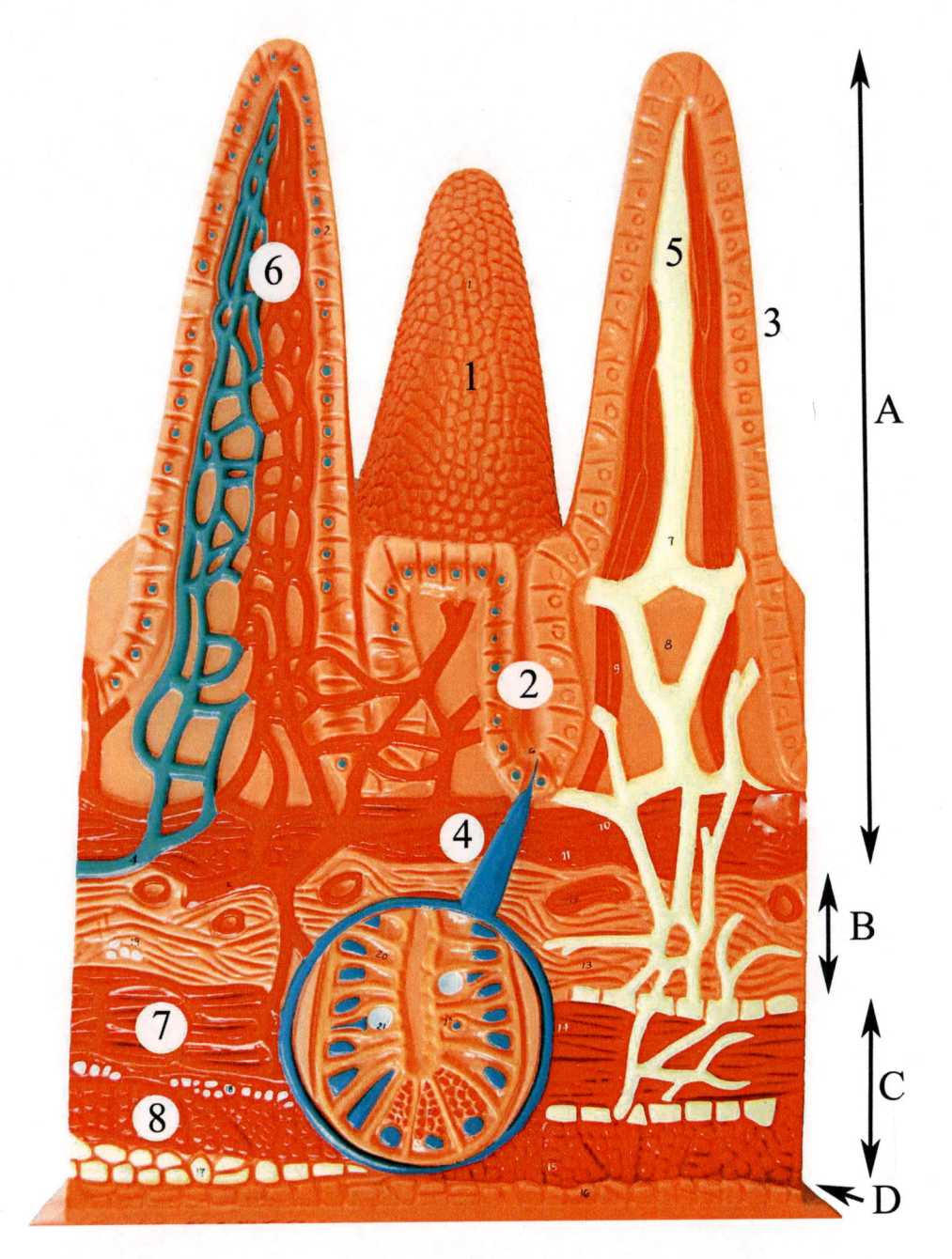

Figure 7-23. Small Intestine with Villi
(Copyright Denoyer-Geppert)

- A. Mucosa
- B. Submucosa
- C. Muscularis externa
- D. Serosa
- 1. Villus
- 2. Intestinal crypt
- 3. Columnar epithelium
- 4. Muscularis mucosae
- 5. Lacteal
- 6. Capillaries
- 7. Circular muscular layer
- 8. Longitudinal muscular layer

Chapter 8
Urinary System Models

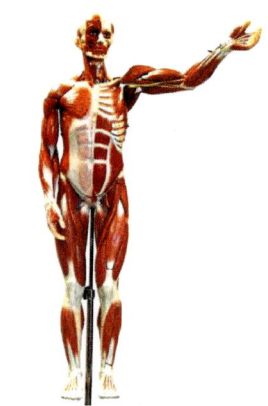

Male Muscle Figure
(Copyright Somso)

Torso
(Copyright 3B Scientific)

(Copyright Somso)

Torso
(Somso)

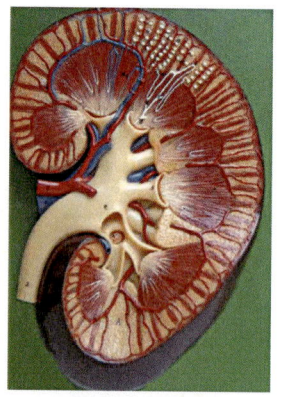

Kidney
(Copyright Somso)

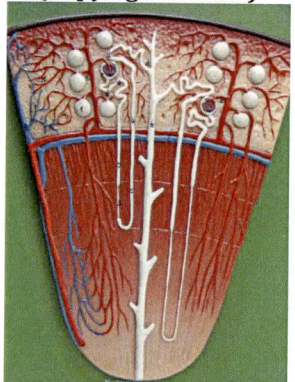

Nephron
(Copyright Somso)

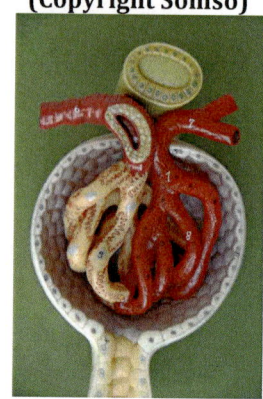

Renal Corpuscle
(Copyright Somso)

Male Muscle Figure

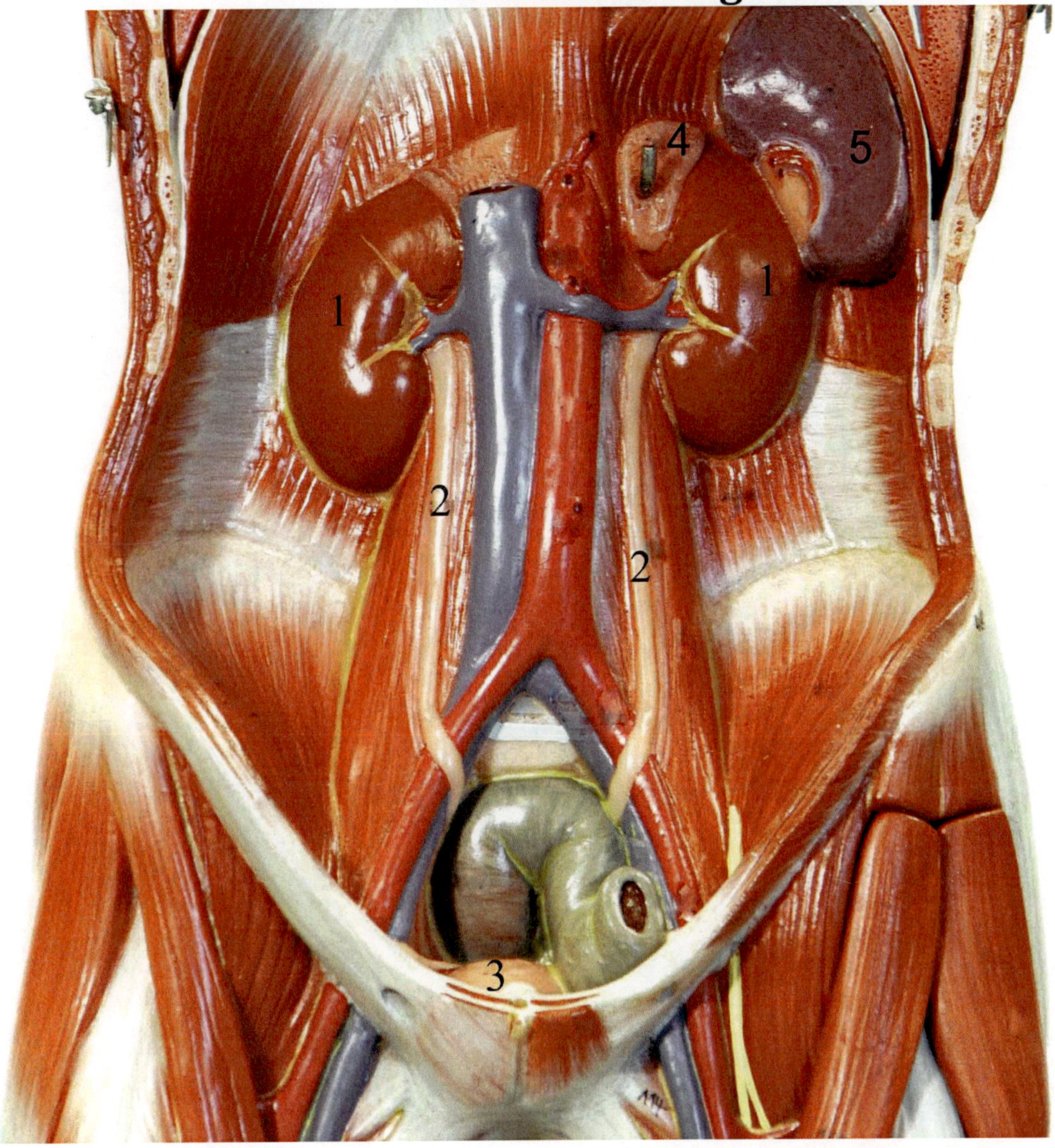

Figure 8-1. Urinary System
(Copyright Somso)
1. Kidney
2. Ureter
3. Urinary bladder
4. Adrenal (suprarenal) gland
5. Spleen

Torso – 3B

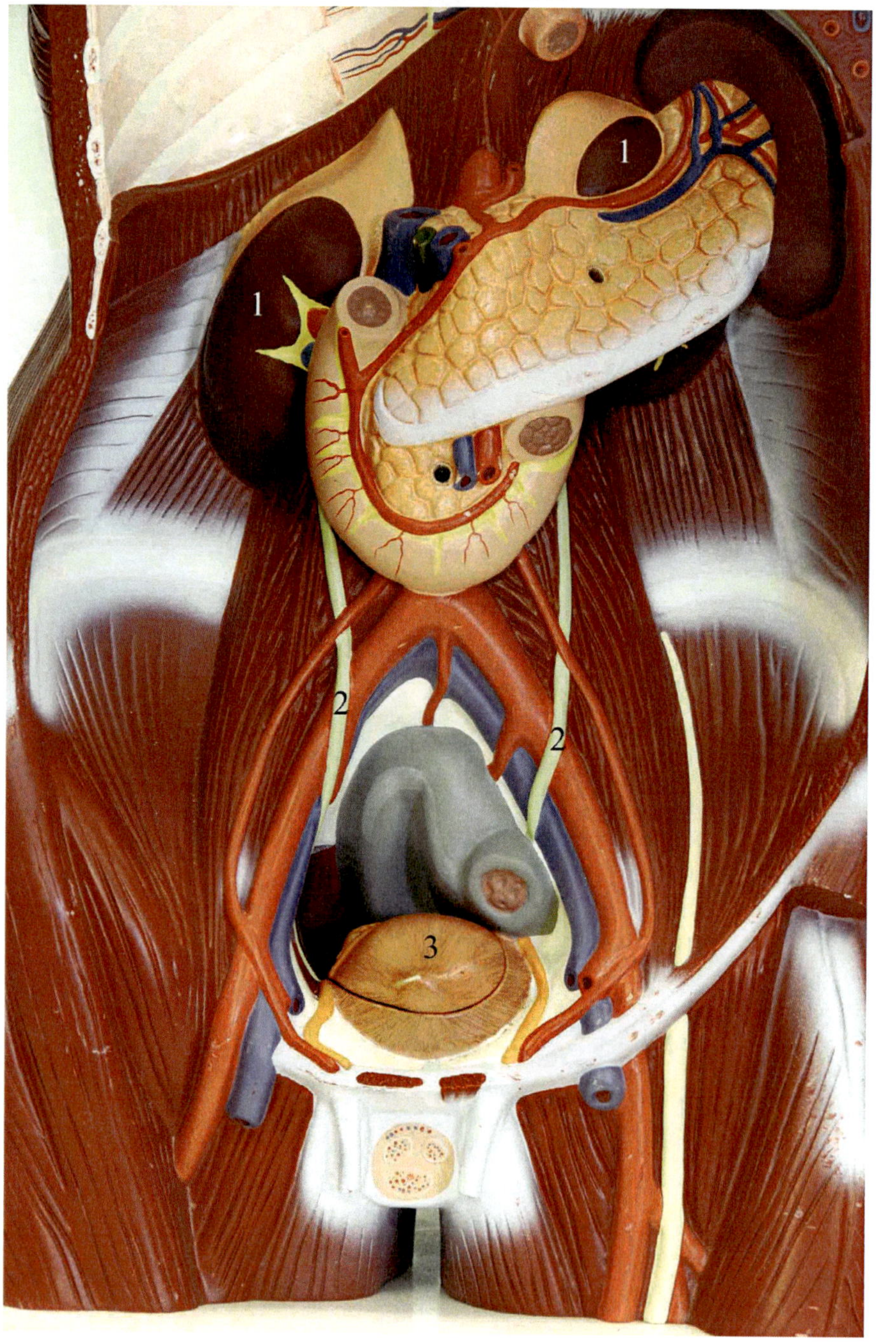

Figure 8-2. Urinary System
(Copyright 3B Scientific)
1. Kidney
2. Ureter
3. Urinary bladder

Torso - Somso

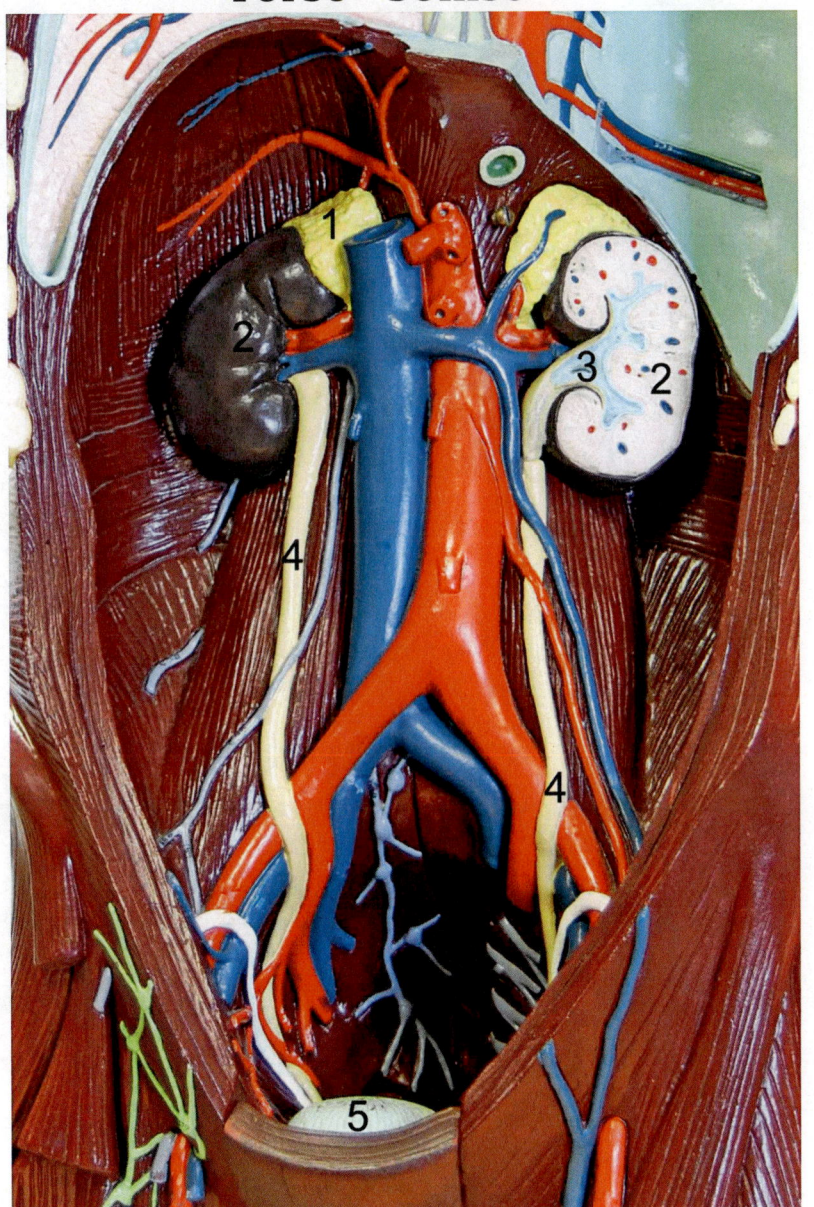

Figure 8-3. Urinary System
(Copyright Somso)
1. Adrenal (suprarenal) gland
2. Kidney
3. Renal pelvis
4. Ureter
5. Urinary Bladder

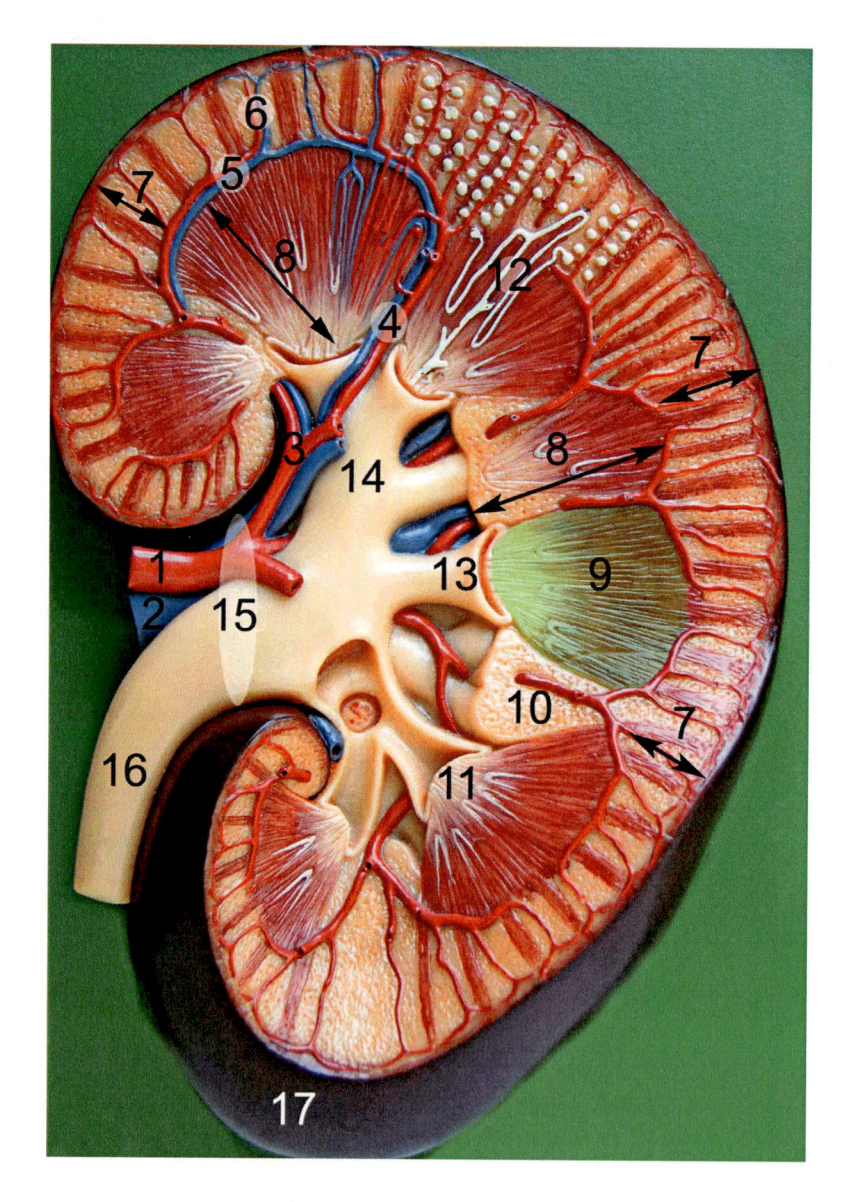

Figure 8-4. Kidney
(Copyright Somso)

1. Renal artery
2. Renal vein
3. Segmental artery
4. Interlobar artery and vein
5. Arcuate artery and vein
6. Cortical radiate (interlobular) artery and vein
7. Renal cortex
8. Renal medulla
9. Renal pyramid
10. Renal column
11. Renal papilla
12. Nephron
13. Minor calyx of renal pelvis
14. Major calyx of renal pelvis
15. Renal hilum
16. Ureter
17. Renal capsule

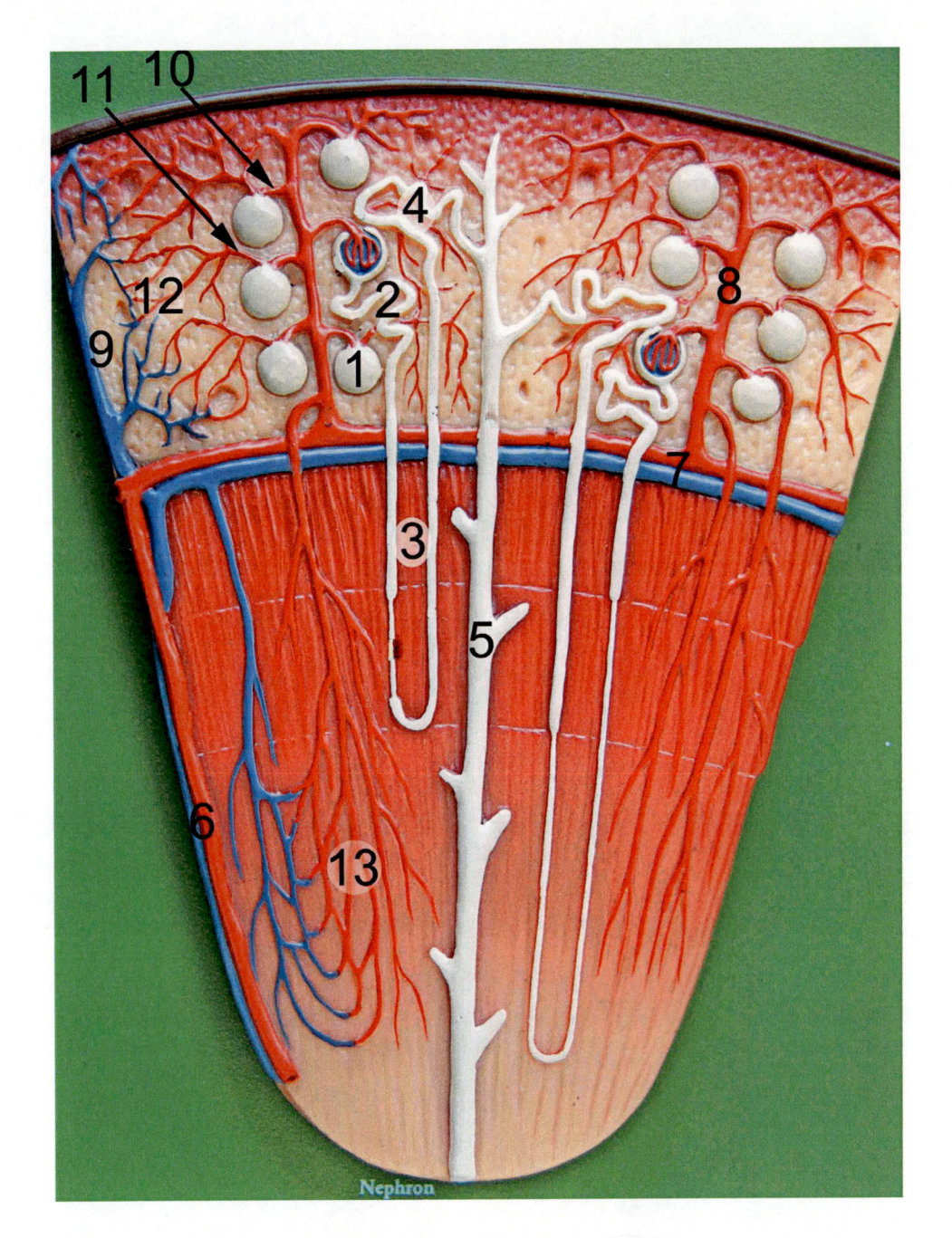

Figure 8-5. Kidney - Nephron
(Copyright Somso)

1. Glomerular (Bowman's) capsule
2. Proximal convoluted tubule
3. Nephron loop (loop of Henle)
4. Distal convoluted tubule
5. Collecting duct
6. Interlobar artery and vein
7. Arcuate artery and vein
8. Cortical radiate (interlobular) artery
9. Cortical radiate (interlobular) vein
10. Afferent arteriole
11. Efferent arteriole
12. Peritubular capillaries
13. Vasa recta

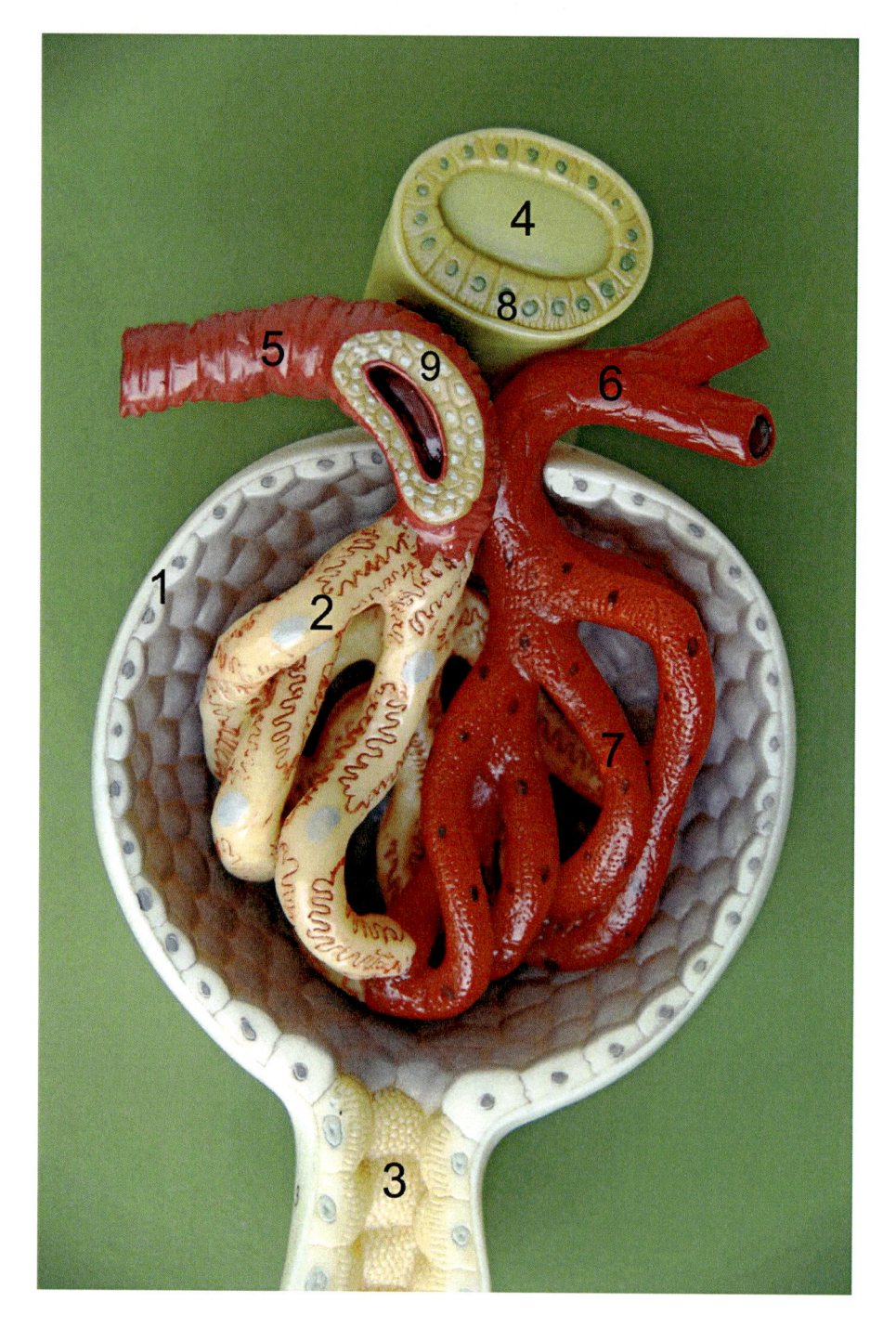

Figure 8-6. Kidney – Renal Corpuscle
(Copyright Somso)

1. Capsular (parietal) epithelium
2. Podocyte (visceral epithelium)
3. Proximal convoluted tubule
4. Distal convoluted tubule
5. Afferent arteriole
6. Efferent arteriole
7. Glomerulus
8. Macula densa
9. Juxtaglomerular cells

Chapter 9
Reproductive System Model

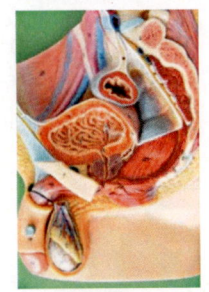

Male
(Copyright Somso)

Torso Female
(Copyright Somso)

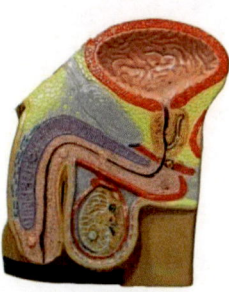

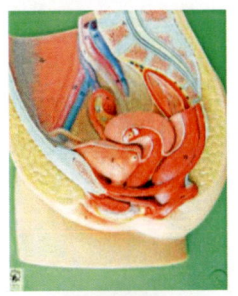

Male
(Purchased from Carolina Biological)

Female
(Copyright Somso)

Male
(Copyright Denoyer-Geppert)

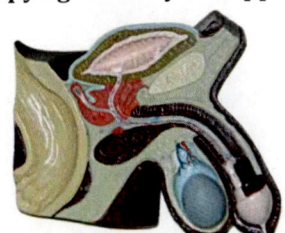

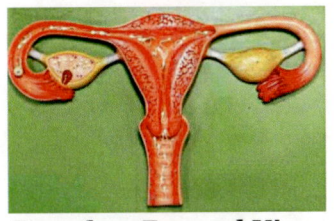

Female – Frontal View
(Copyright Somso)

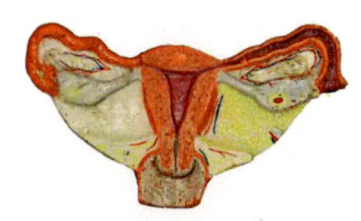

Torso Male
(Copyright Somso)

Female – Frontal view
(Copyright Bobbitt)

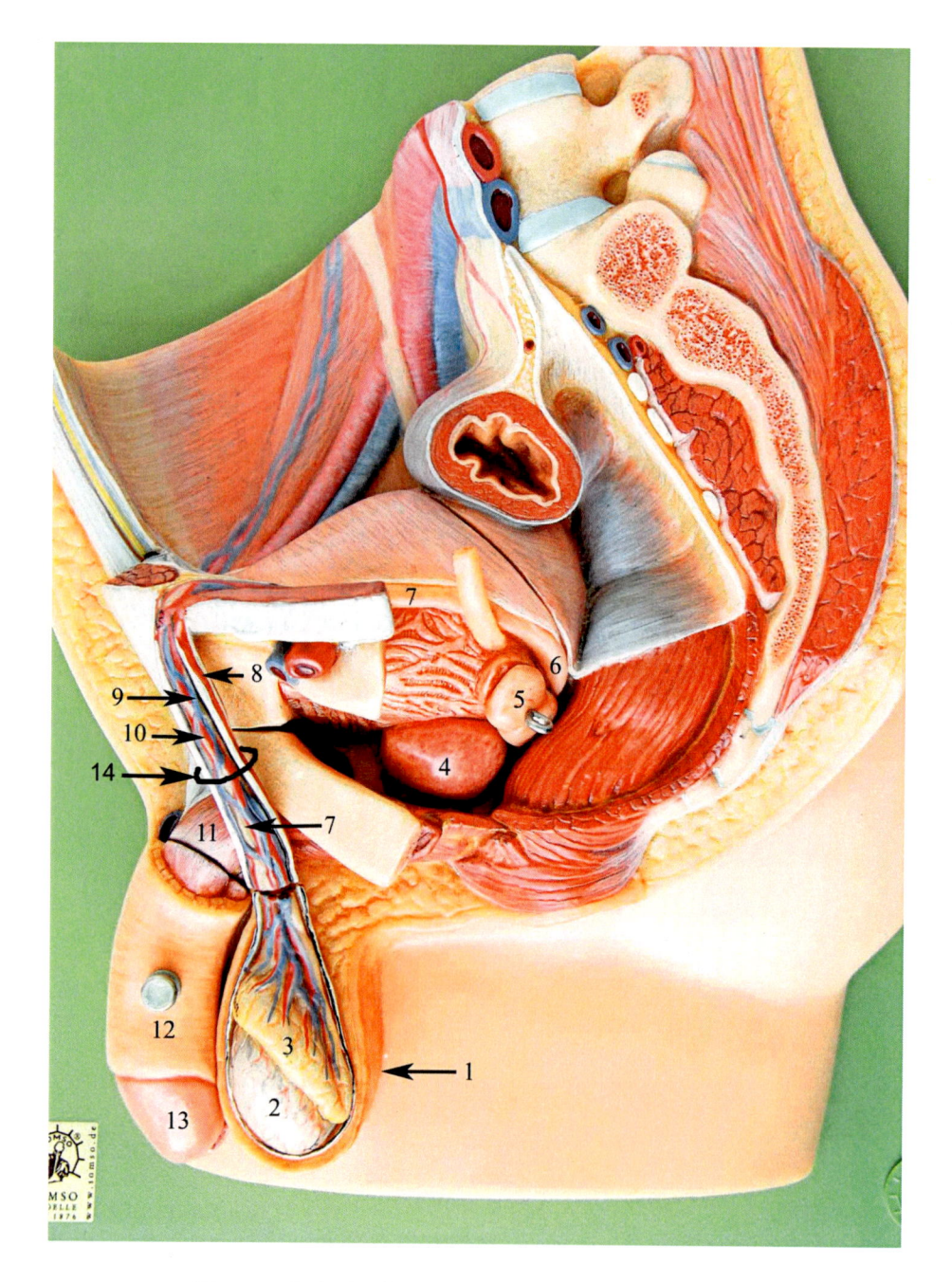

Figure 9-1. Male Reproductive System
(Copyright Somso)

1. Scrotum
2. Testis
3. Epididymis
4. Prostate gland
5. Seminal glands (vesicles)
6. Ampulla of ductus deferens
7. Ductus (vas) deferens
8. Cremaster muscle
9. Pampiniform plexus (blue)
10. Testicular artery (red)
11. Corpus cavernosum
12. Penis
13. Glans of penis
14. Spermatic cord

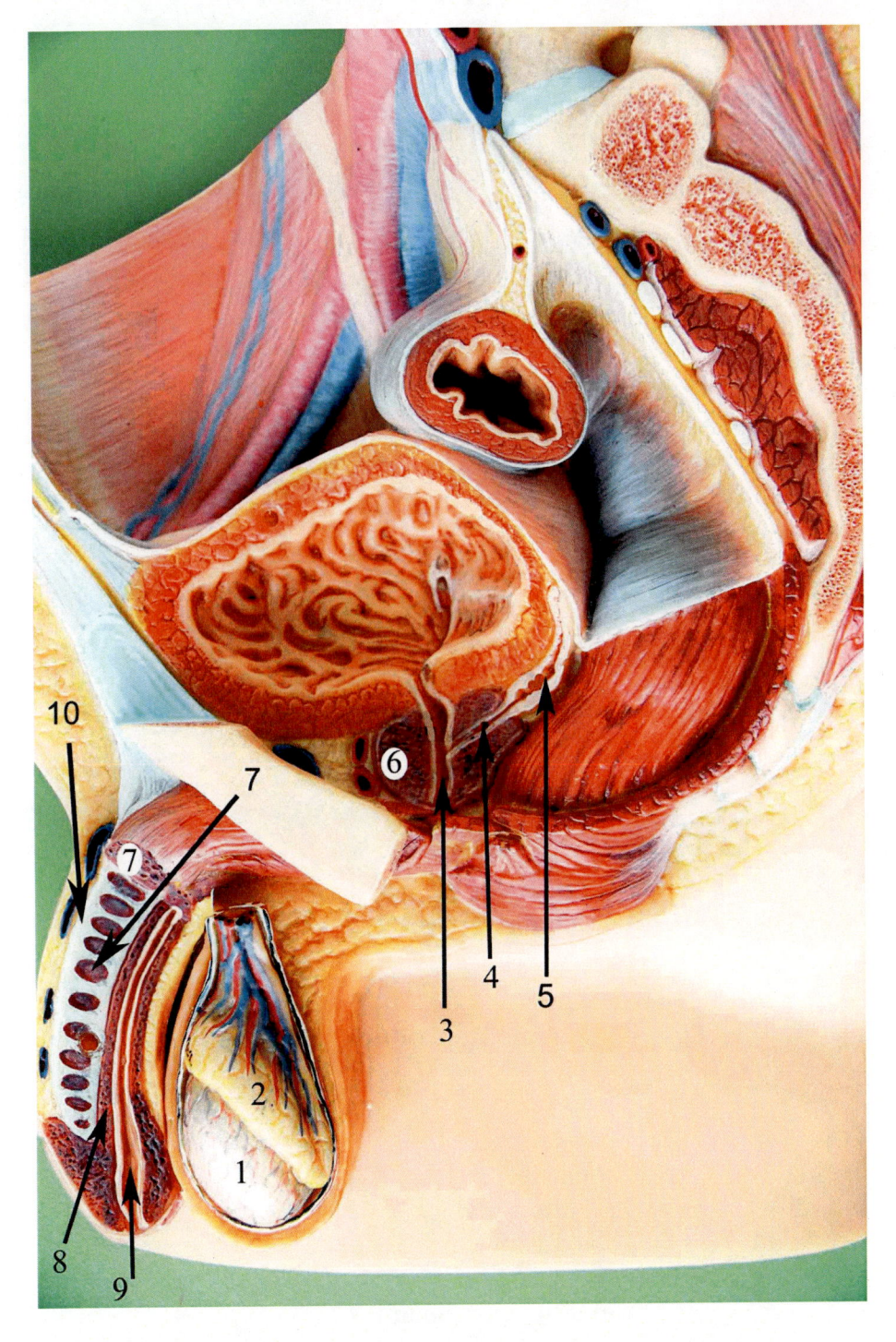

Figure 9-2. Male Reproductive System - interior
(Copyright Somso)

1. Testis
2. Epididymis
3. Prostatic urethra
4. Ejaculatory duct
5. Ampulla of ductus (vas) deferens
6. Prostate gland
7. Corpus cavernosum
8. Corpus spongiosum
9. Spongy urethra
10. Connective tissue septum

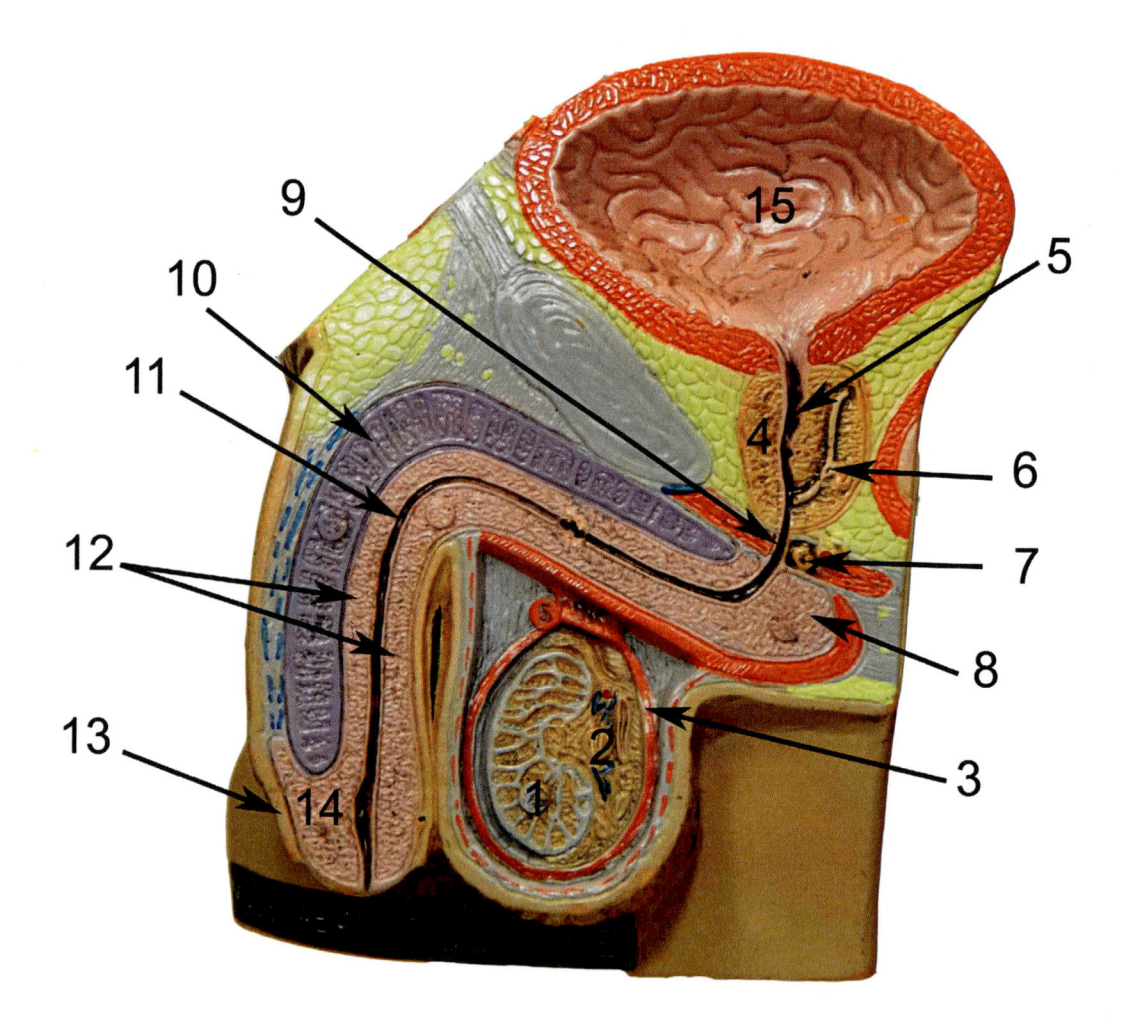

Figure 9-3. Male Reproductive System

1. Testis
2. Epididymis
3. Cremaster muscle
4. Prostate gland
5. Prostatic urethra
6. Ejaculatory duct
7. Bulbo-urethral gland
8. Bulb of corpus spongiosum
9. Membranous urethra
10. Connective tissue septum between corpora cavernosa
11. Spongy urethra
12. Corpus spongiosum
13. Prepuce
14. Glans penis

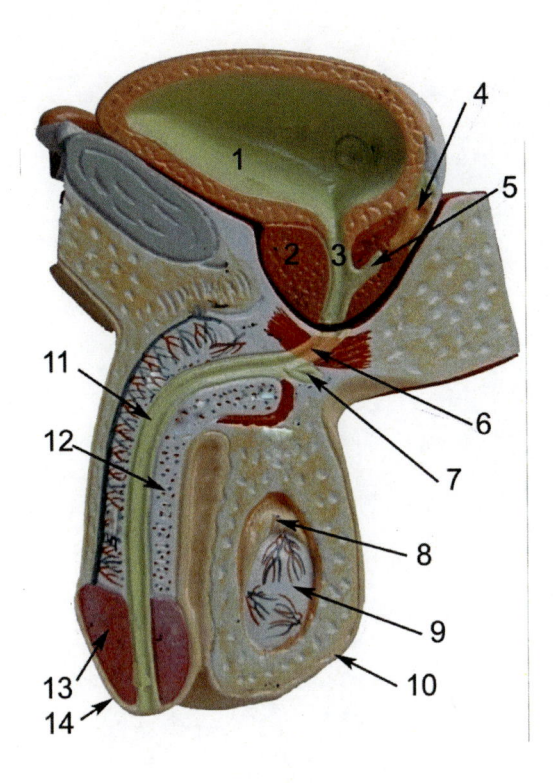

Figure 9-4. Male Reproductive System - medial
(Copyright Denoyer-Geppert)

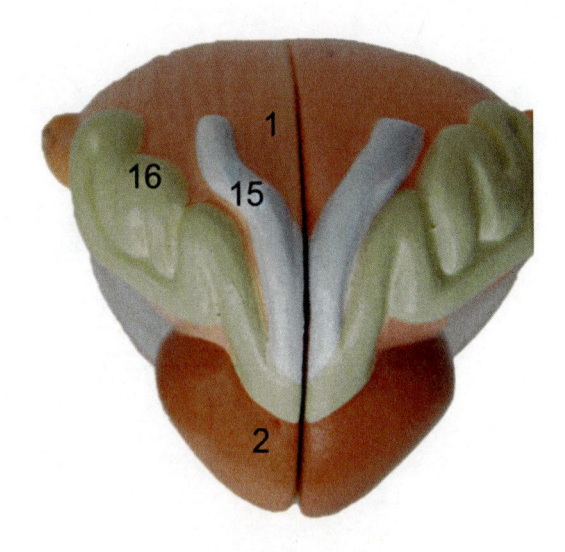

Figure 9-5. Male Reproductive System - posterior

1. Urinary bladder
2. Prostate gland
3. Prostatic urethra
4. Ampulla of ductus deferens
5. Ejaculatory duct
6. Membranous urethra
7. Bulbo-urethral gland
8. Epididymis
9. Testis
10. Scrotum
11. Spongy urethra
12. Corpus spongiosum
13. Glans penis
14. Prepuce
15. Ductus (vas) deferens
16. Seminal glands (vesicles)

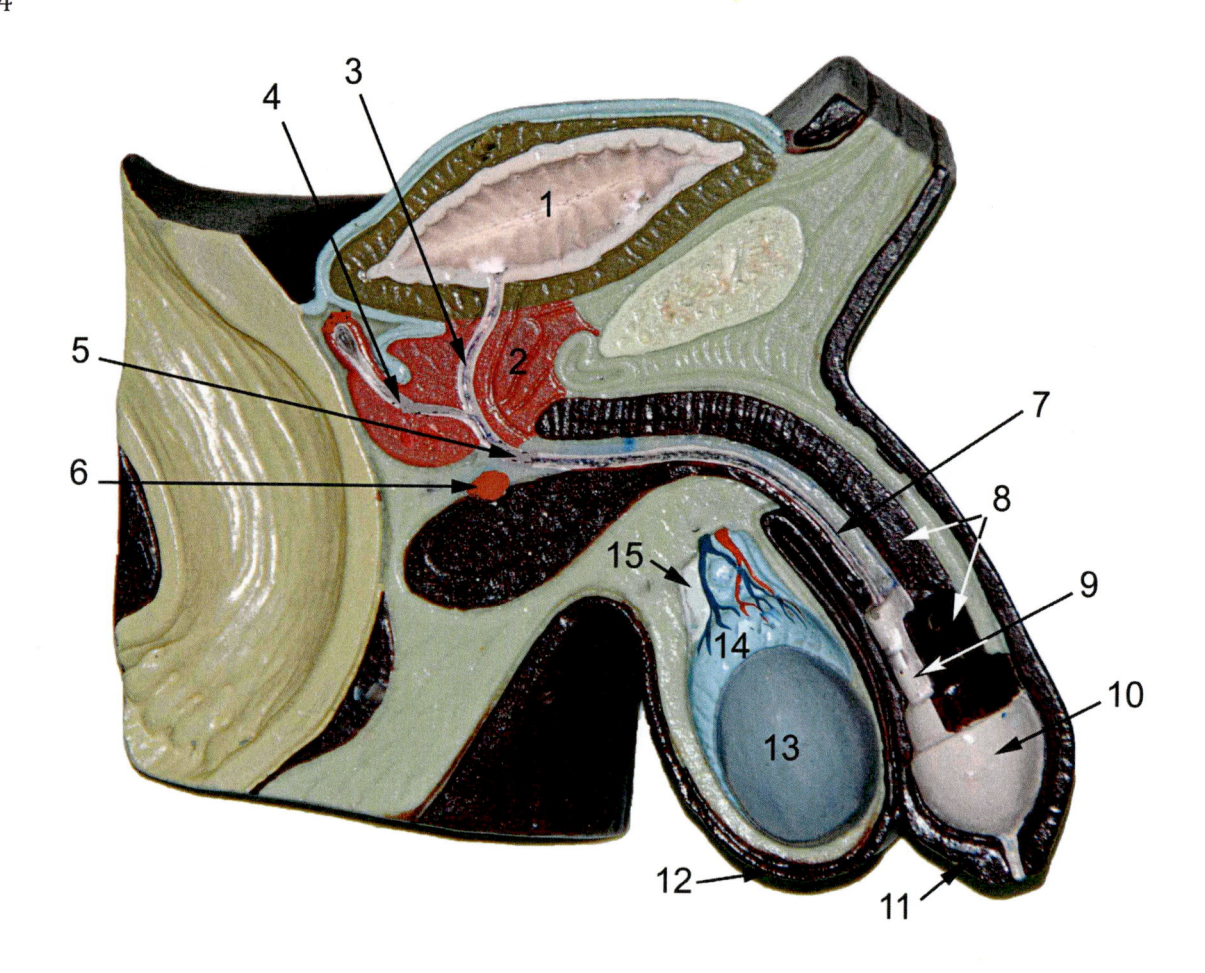

Figure 9-6. Male Reproductive System
(Copyright Somso)

1. Urinary bladder
2. Prostate gland
3. Prostatic urethra
4. Ejaculatory duct
5. Membranous urethra
6. Bulbo-urethral gland
7. Spongy urethra
8. Corpus cavernosum
9. Corpus spongiosum
10. Glans penis
11. Prepuce
12. Scrotum
13. Testis
14. Epididymis
15. Ductus (vas) deferens

don't study

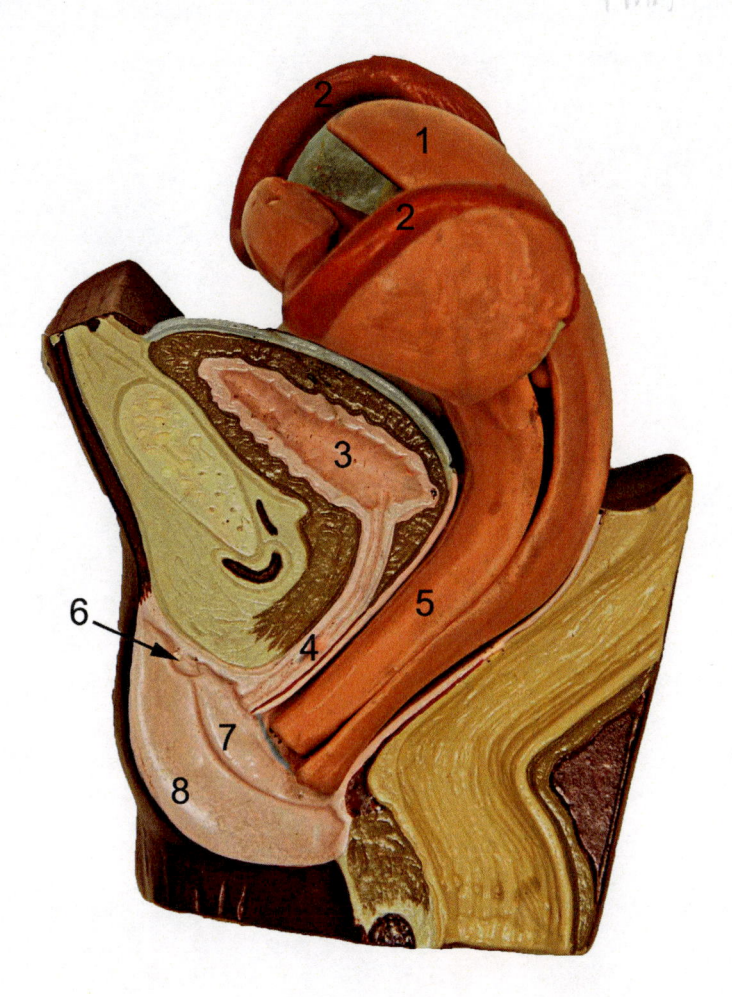

Figure 9-7. Female Reproductive System
(Copyright Somso)

1. Uterus
2. Uterine (Fallopian) tube
3. Urinary bladder
4. Urethra
5. Vagina
6. Clitoris
7. Labium minus
8. Labium majus

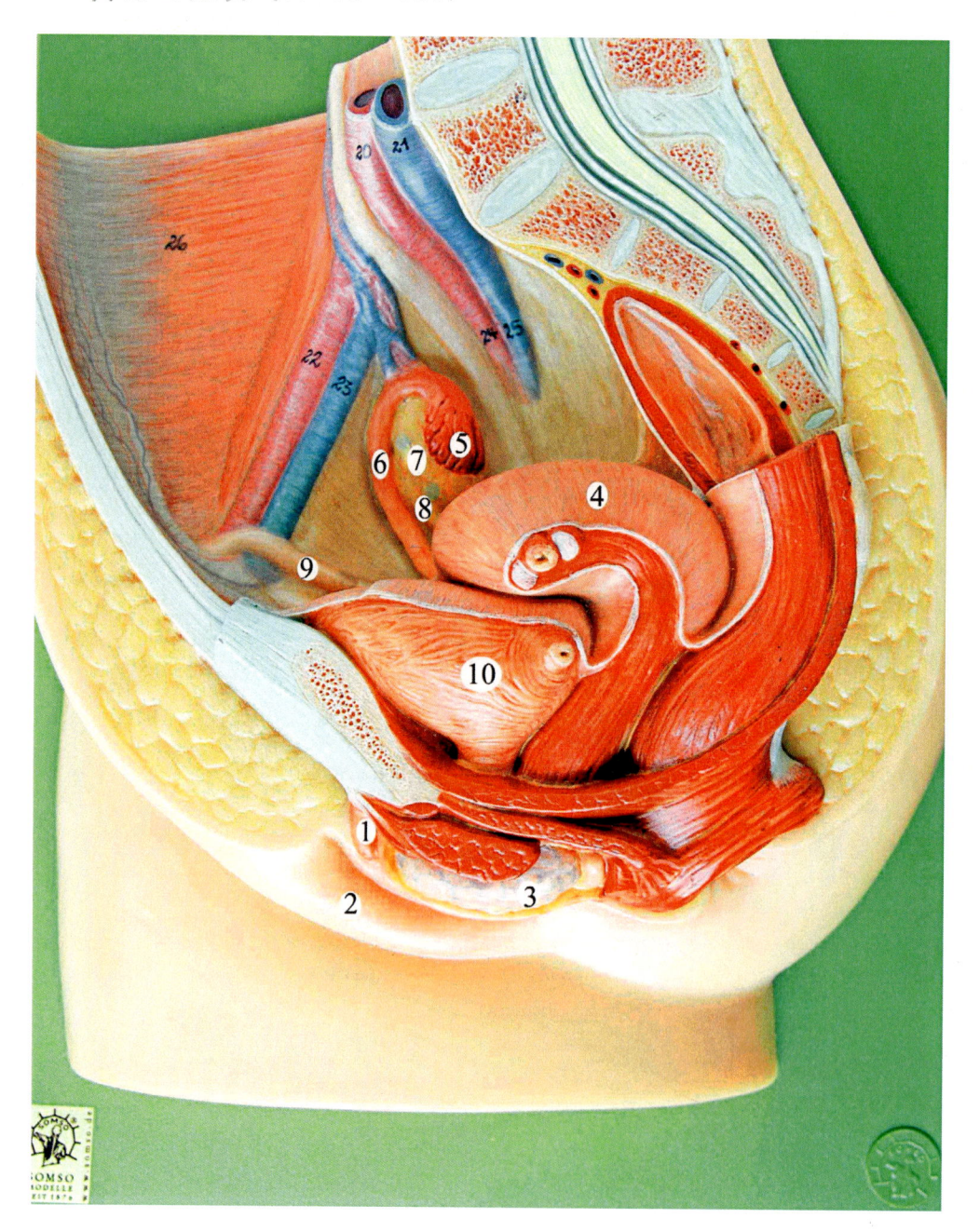

Figure 9-8. Female Reproductive System
(Copyright Somso)

1. Clitoris
2. Labium majus
3. Bulb of vestibule
4. Uterus
5. Fimbriae
6. Uterine (Fallopian) tube
7. Ovary
8. Ovarian ligament
9. Round ligament
10. Urinary bladder

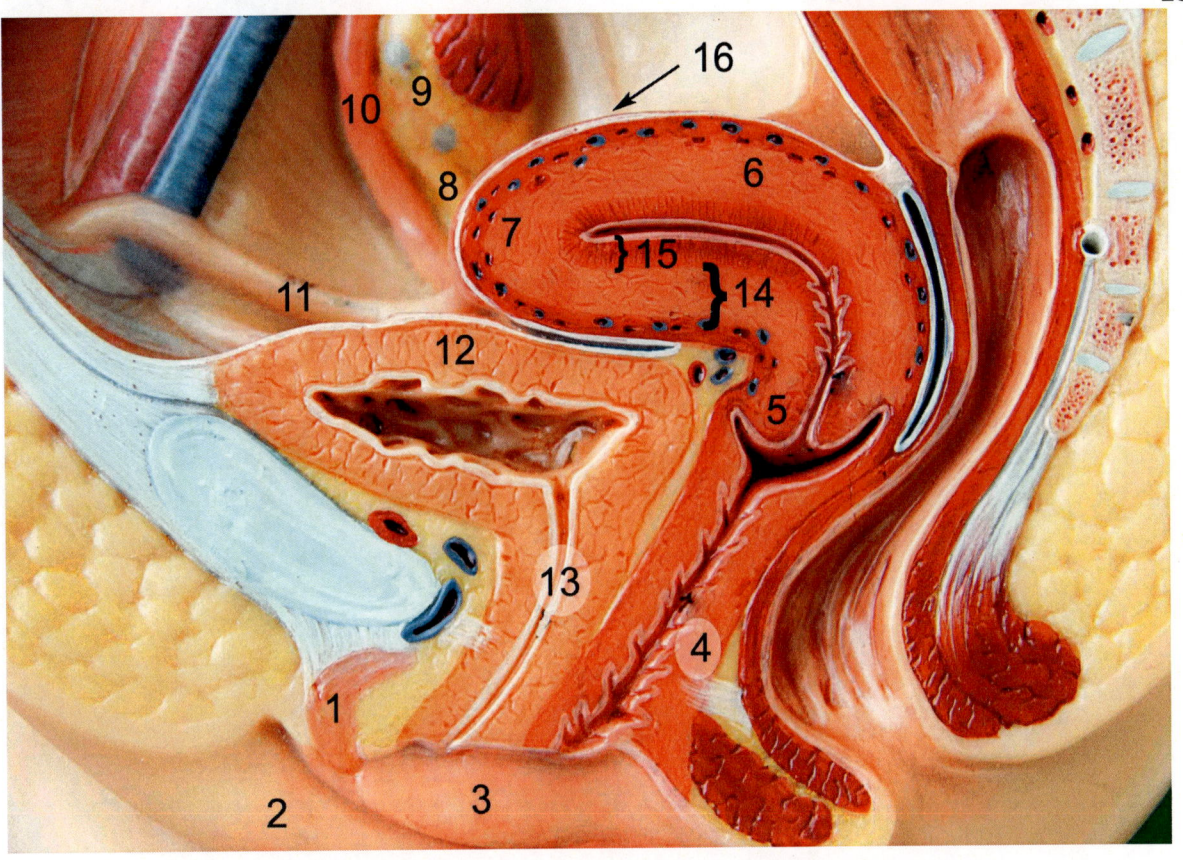

Figure 9-9. Female Reproductive System – interior
(Copyright Somso)

1. Clitoris
2. Labium majus
3. Labium minus
4. Vagina
5. Cervix of uterus
6. Body of uterus
7. Fundus of uterus
8. Ovarian ligament
9. Ovary
10. Uterine (Fallopian) tube
11. Round ligament
12. Urinary bladder
13. Urethra
14. Myometrium
15. Endometrium
16. Perimetrium

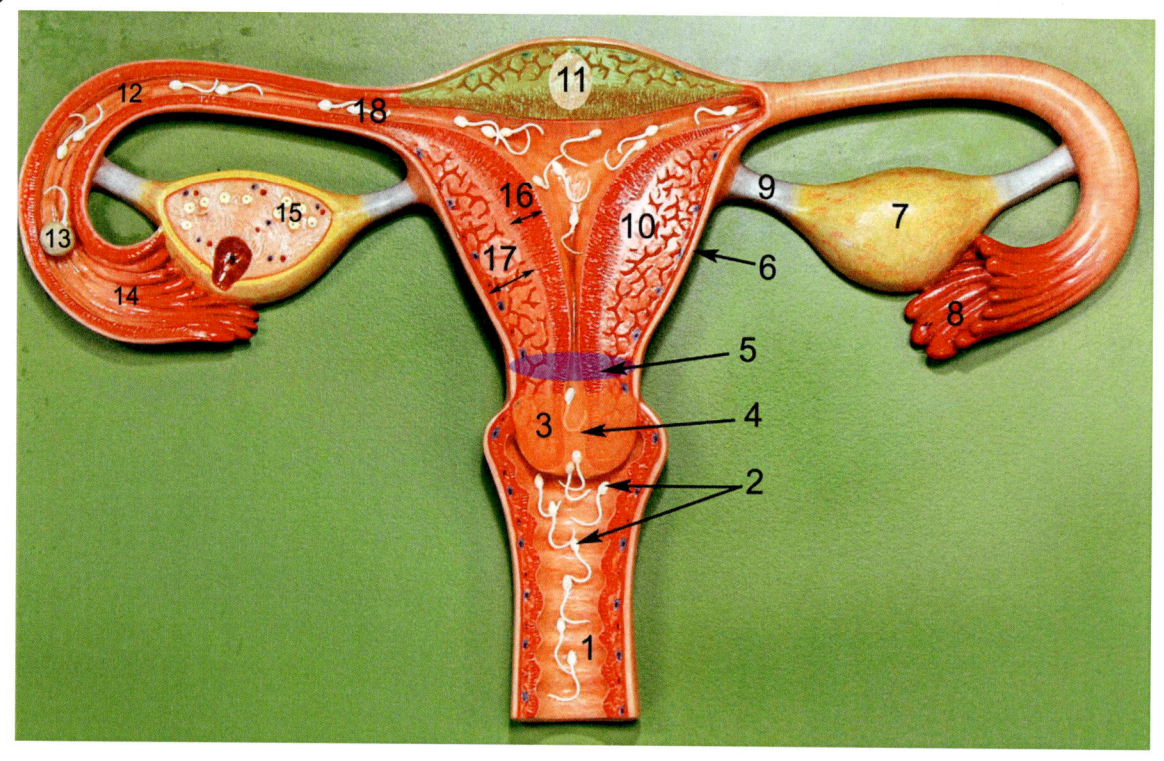

Figure 9-10. Female Reproductive System – fertilization
(Copyright Somso)

1. Vagina
2. Sperm
3. Cervix of uterus
4. Cervical canal
5. Isthmus of uterus
6. Perimetrium
7. Ovary
8. Fimbriae
9. Ovarian ligament
10. Body of uterus
11. Fundus of uterus
12. Ampulla of uterine tube
13. Ovum
14. Infundibulum of uterine tube
15. Primordial follicles
16. Endometrium
17. Myometrium
18. Isthmus of uterine tube

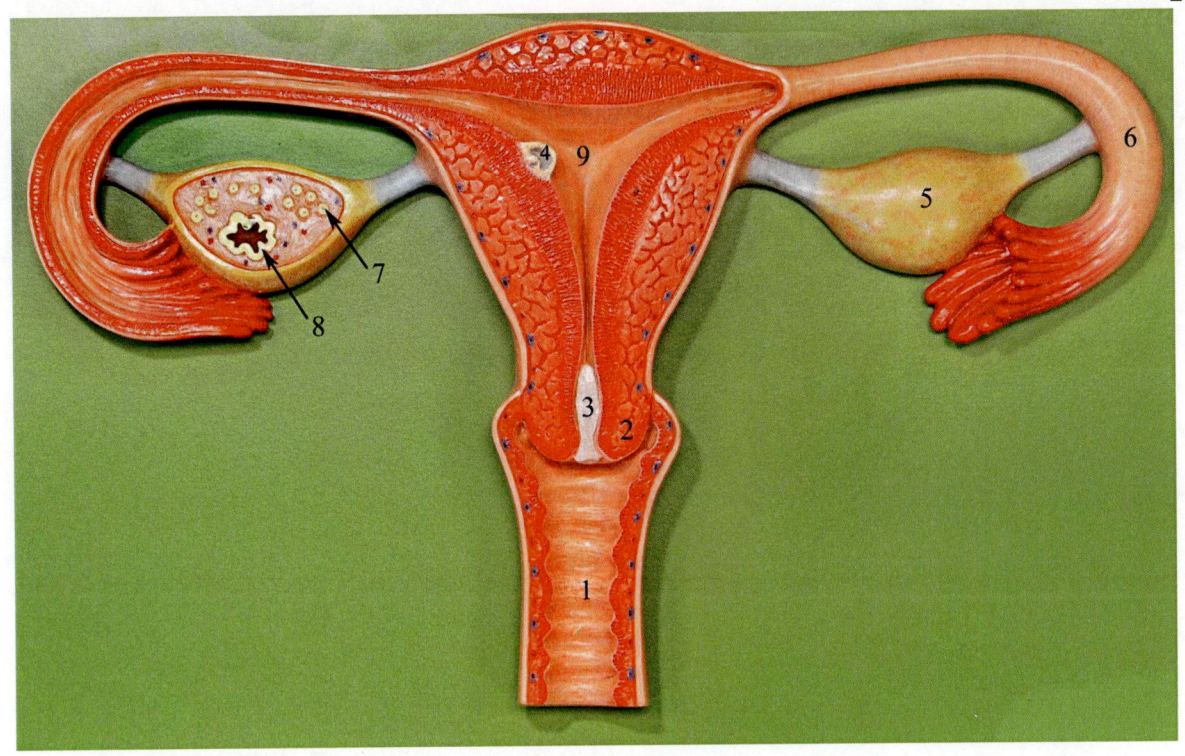

Figure 9-11. Female Reproductive System – embryonic implantation
(Copyright Somso)

1. Vagina
2. Cervix of uterus
3. Cervical (mucous) plug
4. Embryo
5. Ovary
6. Uterine (Fallopian) tube
7. Primordial follicles
8. Corpus luteum
9. Uterus

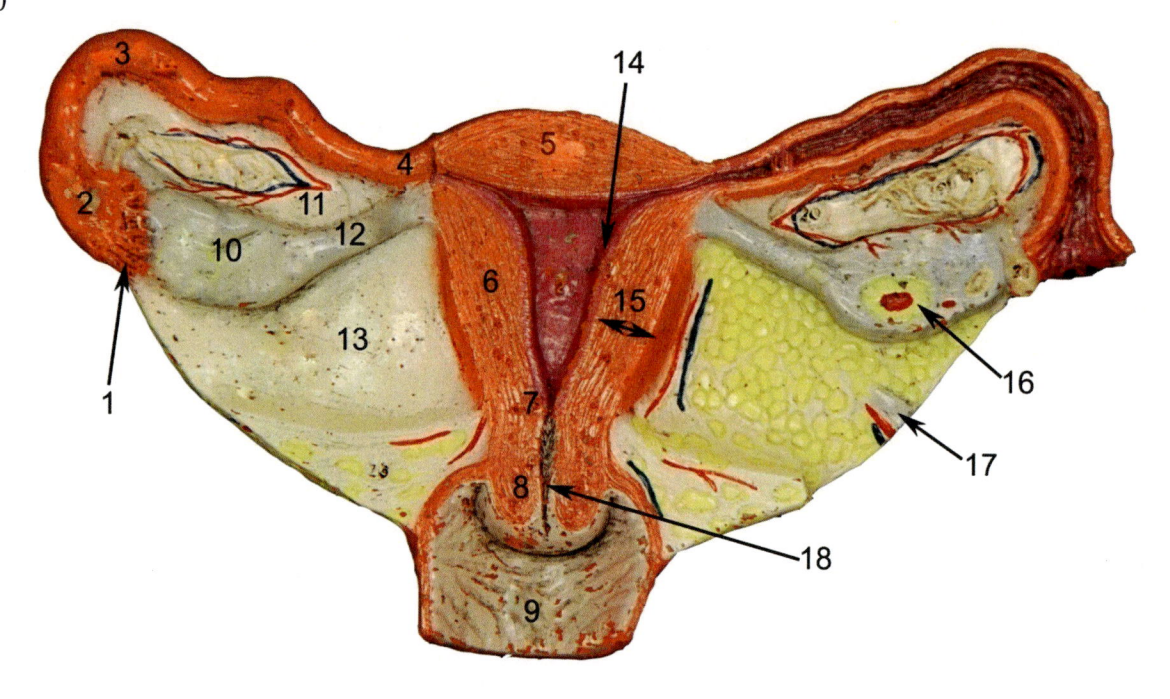

Figure 9-12. Female Reproductive System – frontal view
(Copyright Bobbitt)

1. Fimbria
2. Infundibulum of uterine tube
3. Ampulla of uterine tube
4. Isthmus of uterine tube
5. Fundus of uterus
6. Body of uterus
7. Isthmus of uterus
8. Cervix of uterus
9. Vagina
10. Ovary
11. Mesovarian ligament
12. Ovarian ligament
13. Broad ligament
14. Endometrium
15. Myometrium
16. Corpus luteum in ovary
17. Round ligament
18. Cervical canal

Made in the USA
San Bernardino, CA
19 November 2015